Newnes Radio and Electronics Engineer's Pocket Book
Reviews of 16th Edition

'Everyone involved in electronic engineering should have a copy.'
Electronic Technology

'It cannot be rivalled at the price.'
Radio & Electronics World

'This review copy will never leave my side.'
Electronics Times

'Time to treat yourself.'
What's New in Electronics

'Cannot be too strongly recommended.'
Elektor Electronics

'A fact packed vade mecum for the engineer.'
Electronics & Wireless World

'This really is a handy tome.'
Hi-Fi News

Newnes Radio and Electronics Engineer's Pocket Book

17th edition

Keith Brindley

Heinemann: London

William Heinemann Ltd
10 Upper Grosvenor Street, London W1X 9PA

LONDON MELBOURNE JOHANNESBURG AUCKLAND

First published by George Newnes Ltd, 1940
Thirteenth edition 1962
Fourteenth edition 1972
Fifteenth edition 1978
Reprinted 1979, 1980, 1982 (twice, with additions), 1983
Sixteenth edition 1985
Seventeenth edition first published by
William Heinemann Ltd, 1987

ISBN 0 434 90179 2

Typeset by Vision Typesetting, Manchester
Printed in Great Britain by
Butler & Tanner Ltd, Frome and London

Preface

Radio and electronics reference books are, generally, quite specific in nature; often covering such narrow and detailed aspects that they are of use to only a minority. Those few books which cover more than this tend not to allow easy reference to specific details, and are expensive. My intention in revising this book was to cater for the needs of most people with interests in radio and electronics related areas, while making it easy to locate the required information – at an affordable price. I hope I have succeeded.

My main criterion in choosing what to include and what to discard has been, 'What do *I* look up?' I have tried to include, therefore, *anything* of relevance to radio and electronics referred to in literature. In this respect, a number of tables of units, conversion factors, symbols etc., are newly included. On the other hand, anything for which a calculator is better used, has been discarded.

Keith Brindley

Contents

Abbreviations and symbols

Many abbreviations are found as either capital *or* lower case letters, depending on publishers' styles. Symbols should generally be standard, as shown.

A	Ampere or anode
ABR	Auxiliary bass radiator
a.c.	Alternating current
A/D	Analogue to digital
ADC	Analogue to digital converter
Ae	Aerial
a.f.	Audio frequency
a.f.c.	Automatic frequency control
a.g.c.	Automatic gain control
a.m.	Amplitude modulation
ASA	Acoustical Society of America
ASCII	American Standard Code for Information Interchange
a.t.u.	Aerial tuning unit
AUX	Auxiliary
a.v.c.	Automatic volume control
b	Base of transistor
BAF	Bonded acetate fibre
B & S	Brown & Sharpe (U.S.) wire gauge
b.p.s.	Bits per second
BR	Bass reflex
BSI	British Standards Institution
C	Capacitor, cathode, centigrade, coulomb
c	Collector of transistor, speed of light
CB	Citizen's band
CCD	Charge coupled device
CCIR	International Radio Consultative Committee
CCITT	International Telegraph and Telephone Consultative Committee
CCTV	Closed circuit television
chps	Characters per second
CPU	Central processor unit
CTD	Charge transfer device
CLK	Clock signal
CrO_2	Chromium dioxide
CMOS	Complementary metal oxide semiconductor
c.w.	Continuous wave
D	Diode
d	Drain of an f.e.t.
D/A	Digital to analogue
DAC	Digital to analogue converter
dB	Decibel
d.c.	Direct current
DCC	Double cotton covered
DCE	Data circuit-terminating equipment
DF	Direction finding
DIL	Dual-in-line
DIN	German standards institute
DMA	Direct memory access
DPDT	Double pole, double throw
DPST	Double pole, single throw

DTE	Data terminal equipment
DTL	Diode-transistor logic
DTMF	Dual tone multi-frequency
DX	Long distance reception
e	Emitter of transistor
EAROM	Electrically alterable read only memory
ECL	Emitter coupled logic
e.h.t.	Extremely high tension (voltage)
e.m.f.	Electromotive force
en	Enamelled
EPROM	Erasable programmable read only memory
EQ	Equalisation
ERP	Effective radiated power
EROM	Erasable read only memory
F	Farad, fahrenheit or force
f	Frequency
Fe	Ferrous
FeCr	Ferri-chrome
f.e.t.	Field effect transistor
f.m.	Frequency modulation
f.r.	Frequency response or range
f.s.d.	Full-scale deflection
f.s.k.	Frequency shift keying
G	Giga (10^9)
g	Grid, gravitational constant
H	Henry
h.f.	High frequency
Hz	Hertz (cycles per second)
I	Current
IB	Infinite baffle
i.c.	Integrated circuit
IF	Intermediate frequency
IHF	Institute of High Fidelity (U.S.)
I^2L (IIL)	Integrated injection logic
i.m.d.	Intermodulation distortion
i/p	Input
i.p.s.	Inches per second
k	Kilo (10^3) or cathode
K	Kilo, in computing terms ($= 2^{10} = 1024$), or degrees Kelvin
L	Inductance or lumens
l.e.d.	Light emitting diode
l.f.	Low frequency
LIN	Linear
LOG	Logarithmic
LS	Loudspeaker
LSI	Large scale integration
l.w.	Long wave (approx. 1100–2000 m)
M	Mega (10^6)
m	Milli (10^{-3}) or metres
MHz	Megahertz
m.c.	Moving coil
mic	Microphone
MOS	Metal oxide semiconductor
MPU	Microprocessor unit
MPX	Multiplex
m.w.	Medium wave (approx. 185–560 m)
n	Nano (10^{-9})

NAB	National Association of Broadcasters
Ni-Cad	Nickel-cadmium
n/c	Not connected; normally closed
n/o	Normally open
NMOS	Negative channel metal oxide semiconductor
o/c	Open channel; open circuit
o/p	Output
op-amp	Operational amplifier
p	Pico (10^{-12})
PA	Public address
PABX	Private automatic branch exchange
PAL	Phase alternation, line
p.a.m.	Pulse amplitude modulation
PCB	Printed circuit board
PCM	Pulse code modulation
PLA	Programmable logic array
PLL	Phase locked loop
PMOS	Positive channel metal oxide semiconductor
P.P.M.	Peak programme meter
p.r.f.	Pulse repetition frequency
PROM	Programmable read only memory
PSS	Packet SwitchStream
PSTN	Public Switched Telephone Network
PSU	Power supply unit
PTFE	Polytetrafluoroethylene
PU	Pickup
PUJT	Programmable unijunction transistor
Q	Quality factor; efficiency of tuned circuit, charge
R	Resistance
RAM	Random access memory
RCF	Recommended crossover frequency
RIAA	Record Industry Association of America
r.f.	Radio frequency
r.f.c.	Radio frequency choke (coil)
r.m.s.	Root mean square
ROM	Read only memory
RTL	Resistor transistor logic
R/W	Read/write
RX	Receiver
S	Siemens
s	Source of an f.e.t.
s/c	Short circuit
SCR	Silicon-controlled rectifier
s.h.f.	Super high frequency
SI	International system of units
S/N	Signal-to-noise
SPL	Sound pressure level
SPST	Single pole, single throw
SPDT	Single pole, double throw
SSI	Small scale integration
s.w.	Short wave (approx. 10–60 m)
s.w.g.	Standard wire gauge
s.w.r.	Standing wave ratio
T	Tesla
TDM	Time division multiplex
t.h.d.	Total harmonic distortion
t.i.d.	Transient intermodulation distortion
TR	Transformer

t.r.f.	Tuned radio frequency
TTL	Transistor transistor logic
TTY	Teletype unit
TVI	Television interface; television interference
TX	Transmitter
UART	Universal asynchronous receiver transmitter
u.h.f.	Ultra high frequency (approx. 470–854 MHz)
u.j.t.	Unijunction transistor
ULA	Uncommitted logic array
V	Volts
VA	Volt-amps
v.c.a.	Voltage controlled amplifier
v.c.o.	Voltage controlled oscillator
VCT	Voltage to current transactor
v.h.f.	Very high frequency (approx. 88–216 MHz)
v.l.f.	Very low frequency
VU	Volume unit
W	Watts
Wb	Weber
W/F	Wow and flutter
w.p.m.	Words per minute
X	Reactance
Xtal	Crystal
Z	Impedance
ZD	Zener diode

Letter symbols by unit name

Unit	*Symbol*	*Notes*
ampere	A	SI unit of electric current
ampere (turn)	At	SI unit of magnetomotive force
ampere-hour	Ah	
ampere per metre	$\mathrm{A\,m^{-1}}$	SI unit of magnetic field strength
angstrom	Å	$1\,\text{Å} = 10^{-10}\,\mathrm{m}$
apostilb	asb	$1\,\mathrm{asb} = (1/\pi)\,\mathrm{cd\,m^{-2}}$ A unit of luminance. The SI unit, candela per square metre, is preferred.
atmosphere:		
standard atmosphere	atm	$1\,\mathrm{atm} = 101\,325\,\mathrm{N\,m^{-2}}$
technical atmosphere	at	$1\,\mathrm{at} = 1\,\mathrm{kgf\,cm^{-2}}$
atomic mass unit (unified)	u	The (unified) atomic mass unit is defined as one-twelfth of the mass of an atom of the ^{12}C nuclide. Use of the old atomic mass unit (amu), defined by reference to oxygen, is deprecated.
bar	bar	$1\,\mathrm{bar} = 100\,000\,\mathrm{N\,m^{-2}}$
barn	b	$1\,\mathrm{b} = 10^{-28}\,\mathrm{m^2}$

baud	Bd	Unit of signalling speed equal to one element per second.
becquerel	Bq	1 Bq = 1 s^{-1} SI unit of radioactivity.
bel	B	
bit	b	
British thermal unit	Btu	
calorie (International Table calorie)	cal_{IT}	1 cal°– = 4·1868 J The 9th Conférence Générale des Poids et Mesures adopted the joule as the unit of heat, avoiding the use of the calorie as far as possible.
calorie (thermochemical calorie)	cal	1 cal = 4·1840 J (See note for International Table calorie.)
candela	cd	SI unit of luminous intensity.
candela per square inch	cd in^{-2}	Use of the SI unit, candela per square metre, is preferred.
candela per square metre	cd m^{-2}	SI unit of luminance. The name nit has been used.
candle		The unit of luminous intensity has been given the name *candela*; use of the word *candle* for this purpose is deprecated.
centimetre	cm	
circular mil	cmil	1 cmil = $(\pi/4)\cdot 10^{-6}$ in^2
coulomb	C	SI unit of electrical charge.
cubic centimetre	cm^3	
cubic foot	ft^3	
cubic foot per minute	ft^3 min^{-1}	
cubic foot per second	ft^3 s^{-1}	
cubic inch	in^3	
cubic metre	m^3	
cubic metre per second	m^3 s^{-1}	
cubic yard	yd^3	
curie	Ci	Unit of activity in the field of radiation dosimetry.
cycle	c	
cycle per second	c s^{-1}	Deprecated. Use hertz
decibel	dB	
degree (plane angle)	°	
degree (temperature):		Note that there is no space between the symbol ° and the letter. The use of the word *centigrade* for the Celsius temperature scale was abandoned by the Conférence Générale des Poids et Mesures in 1948.
degree Celsius	°C	
degree Fahrenheit	°F	
degree Kelvin		See Kelvin.
degree Rankine	°R	
dyne	dyn	
electronvolt	eV	

erg	erg	
erlang	E	Unit of telephone traffic.
farad	F	SI unit of capacitance.
foot	ft	
footcandle	fc	Use of the SI unit of illuminance, the lux (lumen per square metre), is preferred.
footlambert	fL	Use of the SI unit, the candela per square metre, is preferred.
foot per minute	$\mathrm{ft\,min^{-1}}$	
foot per second	$\mathrm{ft\,s^{-1}}$	
foot per second squared	$\mathrm{ft\,s^{-2}}$	
foot pound-force	$\mathrm{ft\,lb_f}$	
gal	Gal	$1\ \mathrm{Gal} = 1\ \mathrm{cm\,s^{-2}}$
gallon	gal	The gallon, quart, and pint differ in the US and the UK, and their use is deprecated.
gauss	G	The gauss is the electromagnetic CGS (Centimetre Gram Second) unit of magnetic flux density. The SI unit, tesla, is preferred.
gigaelectronvolt	GeV	
gigahertz	GHz	
gilbert	Gb	The gilbert is the electromagnetic CGS (Centimetre Gram Second) unit of magnetomotive force. Use of the SI unit, the ampere (or ampere-turn), is preferred.
grain	gr	
gram	g	
gray	Gy	$1\ \mathrm{Gy} = 1\ \mathrm{J\,kg^{-1}}$ SI unit of absorbed dose.
henry	H	
hertz	Hz	SI unit of frequency.
horsepower	hp	Use of the SI unit, the watt, is preferred.
hour	h	Time may be designated as in the following example: $9^h46^m30^s$.
inch	in	
inch per second	$\mathrm{in\,s^{-1}}$	
joule	J	SI unit of energy.
joule per Kelvin	$\mathrm{J\,K^{-1}}$	SI unit of heat capacity and entropy.
Kelvin	K	SI unit of temperature (formerly called *degree Kelvin*). The symbol K is now used without the symbol °.
kiloelectronvolt	KeV	

kilogauss	kG	
kilogram	kg	SI unit of mass.
kilogram-force	kg_f	In some countries the name *kilopond* (kp) has been adopted for this unit.
kilohertz	kHz	
kilojoule	kJ	
kilohm	kΩ	
kilometer	km	
kilometer per hour	$km\,h^{-1}$	
kilopond	kp	See kilogram-force.
kilovar	kvar	
kilovolt	kV	
kilovoltampere	kVA	
kilowatt	kW	
kilowatthour	kWh	
knot	kn	$1\ kn = 1\ nmi\,h^{-1}$
lambert	L	The lambert is the CGS (Centimetre Gram Second) unit of luminance. The SI unit, candela per square metre, is preferred.
litre	l	
litre per second	$l\,s^{-1}$	
lumen	lm	SI unit of luminous flux.
lumen per square foot	$lm\,ft^{-2}$	Use of the SI unit, the lumen per square metre, is preferred.
lumen per square metre	$lm\,m^{-2}$	SI unit of luminous excitance.
lumen per watt	$lm\,W^{-1}$	SI unit of luminous efficacy.
lumen second	lm s	SI unit of quantity of light.
lux	lx	$1\ lx = 1\ lm\,m^{-2}$ SI unit of illuminance.
maxwell	Mx	The maxwell is the electromagnetic CGS (Centimetre Gram Second) unit of magnetic flux. Use of the SI unit, the weber, is preferred.
megaelectronvolt	MeV	
megahertz	MHz	
megavolt	MV	
megawatt	MW	
megohm	MΩ	
metre	m	SI unit of length.
mho	mho	$1\ mho = 1\ \Omega^{-1} = 1\ S$
microampere	μA	
microbar	μbar	
microfarad	μF	
microgram	μg	
microhenry	μH	
micrometre	μm	
micron		The name *micrometre* (μm) is preferred.
microsecond	μs	
microwatt	μW	
mil	mil	1 mil = 0·001 in.

mile		
nautical	nmi	
statute	mi	
mile per hour	$mi\,h^{-1}$	
milliampere	mA	
millibar	mbar	mb may be used.
milligal	mGal	
milligram	mg	
millihenry	mH	
millilitre	ml	
millimetre	mm	
conventional millimetre of mercury	mm Hg	$1\ \mathrm{mm\,Hg} = 133{\cdot}322\ \mathrm{N\,m^{-2}}$.
millimicron		The name *nanometre* (nm) is preferred.
millisecond	ms	
millivolt	mV	
milliwatt	mW	
minute (plane angle)	...′	
minute (time)	min	Time may be designated as in the following example: $9^h46^m30^s$.
mole	mol	SI unit of amount of substance.
nanoampere	nA	
nanofarad	nF	
nanometre	nm	
nanosecond	ns	
nanowatt	nW	
nautical mile	nmi	
neper	Np	
newton	N	SI unit of force.
newton metre	N m	
newton per square metre	$N\,m^{-2}$	See pascal.
nit	nt	$1\ \mathrm{nt} = 1\ \mathrm{cd\,m^{-2}}$ See candela per square metre.
oersted	Oe	The oersted is the electromagnetic CGS (Centimetre Gram Second) unit of magnetic field strength. Use of the SI unit, the ampere per metre, is preferred.
ohm	Ω	SI unit of electrical resistance.
ounce (avoirdupois)	oz	
pascal	Pa	SI unit of pressure or stress. $1\ \mathrm{Pa} = 1\ \mathrm{N\,m^{-2}}$
picoampere	pA	
picofarad	pF	
picosecond	ps	
picowatt	pW	
pint	pt	The gallon, quart, and pint differ in the US and the UK, and their use is deprecated.

pound	lb	
poundal	pdl	
pound-force	lb_f	
pound-force foot	$lb_f ft$	
pound-force per square inch	$lb_f in^{-2}$	
pound per square inch		Although use of the abbreviation psi is common, it is not recommended. See pound-force per square inch.
quart	qt	The gallon, quart, and pint differ in the US and the UK, and their use is deprecated.
rad	rd	Unit of absorbed dose in the field of radiation dosimetry.
radian	rad	SI unit of plane angle.
rem	rem	Unit of dose equivalent in the field of radiation dosimetry.
revolution per minute	$r min^{-1}$	Although use of the abbreviation rpm is common, it is not recommended.
revolution per second	$r s^{-1}$	
roentgen	R	Unit of exposure in the field of radiation dosimetry.
second (plane angle)	..."	
second (time)	s	SI unit of time. Time may be designated as in the following example: $9^h 46^m 30^s$.
siemens	S	SI unit of conductance. $1\ S = 1\ \Omega^{-1}$
square foot	ft^2	
square inch	in^2	
square metre	m^2	
square yard	yd^2	
steradian	sr	SI unit of solid angle.
stilb	sb	$1\ sb = 1\ cd\,cm^{-2}$ A CGS unit of luminance. Use of the SI unit, the candela per square metre, is preferred.
tesla	T	SI unit of magnetic flux density. $1\ T = 1\ Wb\,m^{-2}$.
tonne	t	1 t = 1000 kg.
(unified) atomic mass unit	u	See atomic mass unit (unified).
var	var	Unit of reactive power.
volt	V	SI unit of electromotive force.
voltampere	VA	SI unit of apparent power.
watt	W	SI unit of power.
watthour	Wh	
watt per steradian	$W sr^{-1}$	SI unit of radiant intensity.
watt per steradian square metre	$W (sr\,m^2)^{-1}$	SI unit of radiance.
weber	Wb	SI unit of magnetic flux. 1 Wb = 1 V s.
yard	yd	

Electric quantities

Quantity	*Symbol*	*Unit*	*Symbol*
Admittance	Y	siemens	S
Angular frequency	ω	hertz	Hz
Apparent power	S	watt	W
Capacitance	C	farad	F
Charge	Q	coulomb	C
Charge density	ρ	coulomb per square metre	Cm^{-2}
Conductance	G	siemens	S
Conductivity	κ,γ,σ	siemens per metre	Sm^{-1}
Current	I	ampere	A
Current density	j,J	ampere per square metre	Am^{-2}
Displacement	D	coulomb per square metre	Cm^{-2}
Electromotive force	E	volt	V
Energy	E	joule	J
Faraday constant	F	coloumb per mole	$Cmol^{-1}$
Field strength	E	volt per metre	Vm^{-1}
Flux	ψ	coulomb	C
Frequency	v,f	hertz	Hz
Impedance	Z	ohm	Ω
Light, velocity of in a vacuum	c	metre per second	ms^{-1}
Period	T	second	s
Permeability	μ	henry per metre	Hm^{-1}
Permeability of space	μ_o	henry per metre	Hm^{-1}
Permeance	Λ	henry	H
Permittivity	ε	farad per metre	Fm^{-1}
Permittivity of space	ε_o	farad per metre	Fm^{-1}
Phase	φ	—	—
Potential	V,U	volt	V
Power	P	watt	W
Quality factor	Q	—	—
Reactance	X	ohm	Ω
Reactive power	Q	watt	W
Relative permeability	μ_r	—	—
Relative permittivity	ε_r	—	—
Relaxation time	τ	second	s
Reluctance	R	reciprocal henry	H^{-1}
Resistance	R	ohm	Ω
Resistivity	ρ	ohm metre	Ωm
Susceptance	B	siemens	S
Thermodynamic temperature	T	kelvin	K
Time constant	τ	second	s
Wavelength	λ	metre	m

Fundamental constants

Constant	*Symbol*	*Value*
Boltzmann constant	k	$1{\cdot}38062 \times 10^{-23}$ JK^{-1}
Electron charge, proton charge	e	$\pm 1{\cdot}60219 \times 10^{-19}$ C
Electron charge-to-mass ratio	e/m	$1{\cdot}7588 \times 10^{11}$ Ckg^{-1}
Electron mass	m_e	$9{\cdot}10956 \times 10^{-31}$ kg
Electron radius	r_e	$2{\cdot}81794 \times 10^{-15}$ m
Faraday constant	F	$9{\cdot}64867 \times 10^{4}$ Cmol^{-1}
Neutron mass	m_n	$1{\cdot}67492 \times 10^{-27}$ kg
Permeability of space	μ_o	$4\pi \times 10^{-7}$ Hm^{-1}
Permittivity of space	ε_o	$8{\cdot}85419 \times 10^{-12}$ Fm^{-1}
Planck constant	h	$6{\cdot}6262 \times 10^{-34}$ Js
Proton mass	m_p	$1{\cdot}67251 \times 10^{-27}$ kg
Velocity of light	c	$2{\cdot}99793 \times 10^{8}$ ms^{-1}

Electrical relationships

Amperes × ohms = **volts**
Volts ÷ amperes = **ohms**
Volts ÷ ohms = **amperes**
Amperes × volts = **watts**
(Amperes)^2 × ohms = **watts**
(Volts)^2 ÷ ohms = **watts**
Joules per second = **watts**
Coulombs per second = **amperes**
Amperes × seconds = **coulombs**
Farads × volts = **coulombs**
Coulombs ÷ volts = **farads**
Coulombs ÷ farads = **volts**
Volts × coulombs = **joules**
Farads × (volts)^2 = **joules**

Dimensions of physical properties

Length: metre [L]. Mass: kilogram [M]. Time: second [T]. Quantity of electricity: coulomb [Q]. Area: square metre [L^2]. Volume: cubic metre [L^3].

Velocity: metre per second	[LT^{-1}]
Acceleration: metre per second^2	[LT^{-2}]
Force: newton	[MLT^{-2}]
Work: joule	[ML^2T^{-2}]
Power: watt	[ML^2T^{-3}]
Electric current: ampere	[QT^{-1}]
Voltage: volt	[$ML^2T^{-2}Q^{-1}$]
Electric resistance: ohm	[$ML^2T^{-1}Q^{-2}$]
Electric conductance: siemens	[$M^{-1}L^{-2}TQ^2$]

Inductance: henry	(ML^2Q^{-2}]
Capacitance: farad	[$M^{-1}L^{-2}T^2Q^2$]
Current density: ampere per metre^2	[$L^{-2}T^{-1}Q$]
Electric field strength: volt per metre	[$MLT^{-2}Q^{-1}$]
Magnetic flux: weber	[$MLT^2T^{-1}Q^{-1}$]
Magnetic flux density: tesla	[$MT^{-1}Q^{-1}$]
Energy: joule	[ML^2T^{-2}]
Frequency: hertz	[T^{-1}]
Pressure: pascal	[$ML^{-1}T^{-2}$]

Fundamental units

Quantity	*Unit*	*Symbol*
Amount of a substance	mole	mol
Charge	coulomb	C
Length	metre	m
Luminous intensity	candela	cd
Mass	kilogram	kg
Plane angle	radian	rad
Solid angle	steradian	sr
Thermodynamic temperature	kelvin	K
Time	second	s

Greek alphabet

Capital letters	Small letters	Greek name	English equivalent	Capital letters	Small letters	Greek name	English equivalent
Α	α	*Alpha*	a	Ν	ν	*Nu*	n
Β	β	*Beta*	b	Ξ	ξ	*Xi*	x
Γ	γ	*Gamma*	g	Ο	ο	*Omicron*	ŏ
Δ	δ	*Delta*	d	Π	π	*Pi*	p
Ε	ε	*Epsilon*	e	Ρ	ρ	*Rho*	r
Ζ	ζ	*Zeta*	z	Σ	ς	*Sigma*	s
Η	η	*Eta*	é	Τ	τ	*Tau*	t
Θ	θ	*Theta*	th	Υ	υ	*Upsilon*	u
Ι	ι	*Iota*	i	Φ	φ	*Phi*	ph
Κ	κ	*Kappa*	k	Χ	χ	*Chi*	ch
Λ	λ	*Lambda*	l	Ψ	ψ	*Psi*	ps
Μ	μ	*Mu*	m	Ω	ω	*Omega*	ö

Standard units

Ampere Unit of electric current, the constant current which, if maintained in two straight parallel conductors of infinite length of negligible circular cross-section and placed one metre apart in a vacuum, will produce between them a force equal to 2×10^{-7} newton per metre length.

Ampere-hour Unit of quantity of electricity equal to 3,600 coulombs. One unit is represented by one ampere flowing for one hour.

Coulomb Unit of electric charge, the quantity of electricity transported in one second by one ampere.

Farad Unit of electric capacitance. The capacitance of a capacitor between the plates of which there appears a difference of potential of one volt when it is charged by one coulomb of electricity. Practical units are the microfarad (10^{-6} farad), the nanofarad (10^{-9}) and the picofarad (10^{-12} farad).

Henry Unit of electrical inductance. The inductance of a closed circuit in which an electromotive force of one volt is produced when the electric current in the circuit varies uniformly at the rate of one ampere per second. Practical units are the microhenry (10^{-6} henry) and the millihenry (10^{-3} henry).

Hertz Unit of frequency. The number of repetitions of a regular occurrence in one second.

Joule Unit of energy, including work and quantity of heat. The work done when the point of application of a force of one newton is displaced through a distance of one metre in the direction of the force.

Kilovolt-ampere 1,000 volt-amperes.

Kilowatt 1,000 watts.

Mho Unit of conductance, see Siemens.

Newton Unit of force. That force which, applied to a mass of one kilogram, gives it an acceleration of one metre per second per second.

Ohm Unit of electric resistance. The resistance between two points of a conductor when a constant difference of potential of one volt, applied between these two points, produces in the conductor a current of one ampere.

Siemens Unit of conductance, the reciprocal of the ohm. A body having a resistance of 4 ohms would have a conductance of 0·25 siemens.

Tesla Unit of magnetic flux density, equal to one weber per square metre of circuit area.

Volt Unit of electric potential. The difference of electric potential between two points of a conducting wire carrying a constant current of one ampere, when the power dissipated between these points is equal to one watt.

Volt-ampere The product of the root-mean-square volts and root-mean-square amperes.

Watt Unit of power, equal to one joule per second. Volts times amperes equals watts.

Weber Unit of magnetic flux. The magnetic flux which, linking a circuit of one turn, produces in it an electromotive force of one volt as it is reduced to zero at a uniform rate in one second.

Light, velocity of Light waves travel at 300,000 kilometres per second (approximately). Also the velocity of radio waves.

Sound, velocity of Sound waves travel at 332 metres per second in air (approximately) at sea level.

Decimal multipliers

Prefix	Symbol	Multiplier	Prefix	Symbol	Multiplier	Prefix	Symbol	Multiplier
tera	T	10^{12}	deka	da	10	nano	n	10^{-9}
giga	G	10^{9}	deci	d	10^{-1}	pico	p	10^{-12}
mega	M	10^{6}	centi	c	10^{-2}	femto	f	10^{-15}
kilo	k	10^{3}	milli	m	10^{-3}	atto	a	10^{-18}
hecto	h	10^{2}	micro	μ	10^{-6}			

Useful formulae

Boolean Algebra (laws of)

Absorption:
$$A + (A.B) = A$$
$$A.(A + B) = A$$

Annulment:
$$A + 1 = 1$$
$$A.0 = 0$$

Association:
$$(A + B) + C = A + (B + C)$$
$$(A.B).C = A.(B.C)$$

Commutation:
$$A + B = B + A$$
$$A.B = B.A$$

Complements:
$$A + \bar{A} = 1$$
$$A.\bar{A} = 0$$

De Morgan's:
$$(\overline{A + B}) = \bar{A}.\bar{B}$$
$$(\overline{A.B}) = \bar{A} + \bar{B}$$

Distributive:
$$A.(B + C) = (A.B) + (A.C)$$
$$A + (B.C) = (A + B).(A + C)$$

Double negation:
$$\bar{\bar{A}} = A$$

Identity:
$$A + O = A$$
$$A.1 = A$$

Tautology:
$$A.A = A$$
$$A + A = A$$

Capacitance

The capacitance of a parallel plate capacitor can be found from

$$C = \frac{0{\cdot}885\,KA}{d}$$

C is in picofarads, K is the dielectric constant (air = 1), A is the area of the plate in square cm and d the thickness of the dielectric.

Calculation of overall capacitance with:
Parallel capacitors – $C = C_1 + C_2 + \ldots$

Series capacitors – $\frac{1}{C} = \frac{1}{C_1} + \frac{1}{C_2} + \ldots$

Characteristic impedance

$$\text{(open wire) } Z = 276 \log \frac{2D}{d} \text{ ohms}$$

where D = wire spacing, d = wire diameter } in same units.

$$\text{(coaxial) } Z = \frac{138}{\sqrt{(K)}} \log \frac{d_o}{d_i} \text{ ohms}$$

where K = dielectric constant, d_o = outside diameter of inner conductor, d_i = inside diameter of outer conductor.

Dynamic resistance

In a parallel-tuned circuit at resonance the dynamic resistance is

$$R_d = \frac{L}{Cr} = Q\omega L = \frac{Q}{\omega C} \text{ ohms}$$

where L = inductance (henries), C = capacitance (farads), r = effective series resistance (ohms), Q = Q-value of coil, and $\omega = 2\pi$ × frequency (hertz).

Frequency—wavelength—velocity

(See also Resonance.)
The velocity of propagation of a wave is

$$v = f\lambda \text{ metres per second}$$

where f = frequency (hertz) and λ = wavelength (metres).

For electromagnetic waves in free space the velocity of propagation v is approximately 3×10^8 m/sec, and if f is expressed in kilohertz and λ in metres

$$f = \frac{300{,}000}{\lambda} \text{ kilohertz} \qquad f = \frac{300}{\lambda} \text{ megahertz}$$

or

$$\lambda = \frac{300{,}000}{f} \text{ metres} \qquad \lambda = \frac{300}{f} \text{ metres}$$

f in kilohertz $\qquad$ f in megahertz

Horizon distance

Horizon distance can be calculated from the formula

$$S = 1{\cdot}42\sqrt{H}$$

where S = distance in miles and H = height in feet above sea level.

Impedance
The impedance of a circuit comprising inductance, capacitance and resistance in series is

$$Z = \sqrt{R^2 + \left(\omega L - \frac{1}{\omega C}\right)^2}$$

where R = resistance (ohms), $\omega = 2\pi \times$ frequency (hertz), L = inductance (henries), and C = capacitance (farads).

Inductance
Single layer coils

$$L \text{ (in microhenries)} = \frac{a^2N^2}{9a + 10l} \text{ approximately}$$

If the desired inductance is known, the number of turns required may be determined by the formula

$$N = \frac{5L}{na^2}\left[1 + \sqrt{\left(1 + \frac{0{\cdot}36n^2a^3}{L}\right)}\right]$$

where N = number of turns, a = radius of coil in inches, n = number of turns per inch, L = inductance in microhenries (μH) and l = length of coil in inches.

Calculation of overall inductance with:
Series inductors – $L = L_1 + L_2 + \ldots.$

Parallel inductors – $\frac{1}{L} = \frac{1}{L_1} + \frac{1}{L_2} + \ldots.$

Meter conversions
Increasing range of ammeters or milliammeters
Current range of meter can be increased by connecting a shunt resistance across meter terminals. If R_m is the resistance of the meter; R_s the value of the shunt resistance and n the number of times it is wished to multiply the scale reading, then

$$R_s = \frac{R_m}{(n - 1)}.$$

Increasing range of voltmeters
Voltage range of meter can be increased by connecting resistance in series with it. If this series resistance is R_s and R_m and n as before, then $R_s = R_m \times (n - 1)$.

Negative feedback
Voltage feedback

$$\text{Gain with feedback} = \frac{A}{1 + Ab}$$

where A is the original gain of the amplifier section over which feedback is applied (including the output transformer if included) and b is the fraction of the output voltage fed back.

$$\text{Distortion with feedback} = \frac{d}{1 + Ab} \text{ approximately}$$

where d is the original distortion of the amplifier.

Ohm's Law

$$I = \frac{V}{R} \qquad V = IR \qquad R = \frac{V}{I}$$

where I = current (amperes), V = voltage (volts), and R = resistance (ohms).

Power

In a d.c. circuit the power developed is given by

$$W = VI = \frac{V^2}{R} = I^2R \text{ watts}$$

where V = voltage (volts), I = current (amperes), and R = resistance (ohms).

Power ratio

$$P = 10 \log \frac{P_1}{P_2}$$

where P = ratio in decibels. P_1 and P_2 are the two power levels.

Q

The Q value of an inductance is given by

$$Q = \frac{\omega L}{R}$$

Reactance

The reactance of an inductor and a capacitor respectively is given by

$$X_L = \omega L \text{ ohms} \qquad X_C = \frac{1}{\omega C} \text{ ohms}$$

where $\omega = 2\pi \times$ frequency (hertz), L = inductance (henries), and C = capacitance (farads).
The total resistance of an inductance and a capacitance in series is $X_L - X_C$.

Resistance

Calculation of overall resistance with:
Series resistors – $R = R_1 + R_2 + \ldots.$

Parallel resistors – $$\frac{1}{R} = \frac{1}{R_1} + \frac{1}{R_2} + \ldots.$$

Resonance

The resonant frequency of a tuned circuit is given by

$$f = \frac{1}{2\pi\sqrt{LC}} \text{ hertz}$$

where L = inductance (henries), and C = capacitance (farads).
If L is in microhenries (μH) and C is in picofarads, this becomes—

$$f = \frac{10^6}{2\pi\sqrt{LC}} \text{ kilohertz}$$

The basic formula can be rearranged

$$L = \frac{1}{4\pi^2 f^2 C} \text{ henries} \qquad C = \frac{1}{4\pi^2 f^2 L} \text{ farads.}$$

Since $2\pi f$ is commonly represented by ω, these expressions can be written

$$L = \frac{1}{\omega^2 C} \text{ henries} \qquad C = \frac{1}{\omega^2 L} \text{ farads.}$$

Time constant

For a combination of inductance and resistance in series the time constant (i.e. the time required for the current to reach 63% of its final value) is given by

$$\tau = \frac{L}{R} \text{ seconds}$$

where L = inductance (henries), and R = resistance (ohms).

For a combination of capacitance and resistance in series the time constant (i.e. the time required for the voltage across the capacitance to reach 63% of its final value) is given by

$$\tau = CR \text{ seconds}$$

where C = capacitance (farads), and R = resistance (ohms).

Transformer ratios

The ratio of a transformer refers to the ratio of the number of turns in one winding to the number of turns in the other winding. To avoid confusion it is always desirable to state in which sense the ratio is being expressed: e.g. the 'primary-to-secondary' ratio n_p/n_s. The turns ratio is related to the impedance ratio thus

$$\frac{n_p}{n_s} = \sqrt{\frac{Z_p}{Z_s}}$$

where n_p = number of primary turns, n_s = number of secondary turns, Z_p = impedance of primary (ohms), and Z_s = impedance of secondary (ohms).

Wattage rating

If resistance and current values are known,

$$W = I^2 R \text{ when } I \text{ is in amperes}$$

or

$$W = \frac{\text{Milliamps.}^2}{1{,}000{,}000} \times R.$$

If wattage rating and value of resistance are known, the safe current for the resistor can be calculated from

$$\text{milliamperes} = 1{,}000 \times \sqrt{\frac{\text{Watts}}{\text{Ohms}}}$$

Wavelength of tuned circuit

Formula for the wavelength in metres of a tuned oscillatory circuit is : $1885\sqrt{LC}$, where L = inductance in microhenries and C = capacitance in microfarads.

Resistor and capacitor colour coding

Colour	Band A	Band B	Band C (multiplier): Resistors	Band C (multiplier): Capacitors	Band D (tolerance): Resistors	Band D (tolerance): Capacitors Up to 10 pF
Black	—	0	1	1	—	2 pF
Brown	1	1	10	10	±1%	0·1 pF
Red	2	2	100	100	±2%	—
Orange	3	3	1,000	1,000	—	—
Yellow	4	4	10,000	10,000	—	—
Green	5	5	100,000	—	—	0·5 pF
Blue	6	6	1,000,000	—	—	—
Violet	7	7	10,000,000	—	—	—
Grey	8	8	10^8	0·01 µF	—	0·25 pF
White	9	9	10^9	0^1 µF	—	1 pF
Silver	—	—	0·01	—	±10%	—
Gold	—	—	0·1	—	±5%	—
Pink	—	—	—	—	—	—
None	—	—	—	—	±20%	—

Resistor and capacitor letter and digit code

(BS 1852)

Resistor values are indicated as follows:

0·47 Ω	marked	R47	100 Ω	marked	100R
1 Ω		1R0	1 kΩ		1K0
4·7 Ω		4R7	10 kΩ		10K
47 Ω		47R	10 MΩ		10M

A letter following the value shows the tolerance.
F = ±1%; G = ±2%; J = ±5%; K = ±10%; M = ±20%;
R33M = 0·33Ω ±20%; 6K8F = 6·8 kΩ ±1%.

Capacitor values are indicated as:

0·68 pF	marked	p68	6·8 nf	marked	6n8
6·8 pf		6p8	1000 nF		1µ0
1000 pF		1n0	6·8 µF		6µ8

Tolerance is indicated by letters as for resistors. Values up to 999 pF are marked in pF, from 1000 pf to 999 000 pF (= 999 nF) as nF (1000 pF = 1 nF) and from 1000 nF (= 1 µF) upwards as µF.

Some capacitors are marked with a code denoting the value in pF (first two figures) followed by a multiplier as a power of ten (3 = 10^3). Letters denote tolerance as for resistors but C = ±0·25 pf. E.g. 123 J = 12 pF × 10^3 ± 5% = 12 000 pF (or 0·12 µF).

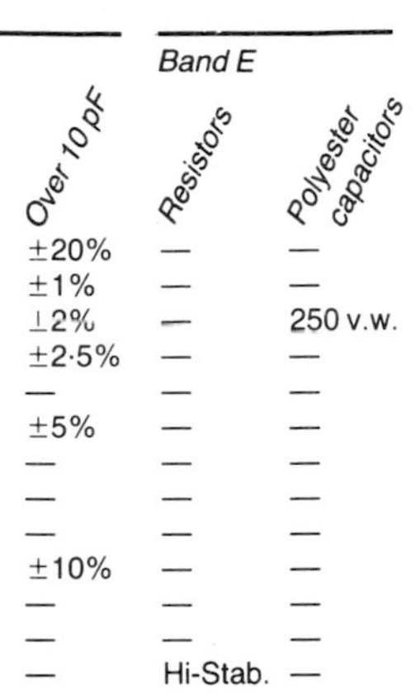

Over 10 pF	Band E Resistors	 Polyester capacitors
±20%	—	—
±1%	—	—
±2%	—	250 v.w.
±2·5%	—	—
—	—	—
±5%	—	—
—	—	—
—	—	—
—	—	—
±10%	—	—
—	—	—
—	—	—
—	Hi-Stab.	—
—	—	—

Note that adjacent bands may be of the same colour unseparated.

Preferred values

E12 Series

1·0 1·2 1·5 1·8 2·2 2·7
3·3 3·9 4·7 5·6 6·8 8·2
and their decades

E24 Series

1·0 1·1 1·2 1·3 1·5 1·6
1·8 2·0 2·2 2·4 2·7 3·0
3·3 3·6 3·9 4·3 4·7 5·1
5·6 6·2 6·8 7·5 8·2 9·1
and their decades

Tantalum capacitors

	1	2	3	4
Black	—	0	× 1	10 V
Brown	1	1	× 10	
Red	2	2	× 100	
Orange	3	3	—	
Yellow	4	4	—	6·3 V
Green	5	5	—	16 V
Blue	6	6	—	20 V
Violet	7	7	—	
Grey	8	8	× 0·01	25 V
White	9	9	× 0·1	3 V
				(Pink 35 V)

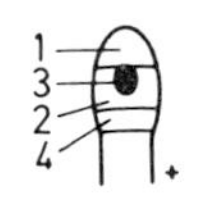

Resistor and capacitor colour coding

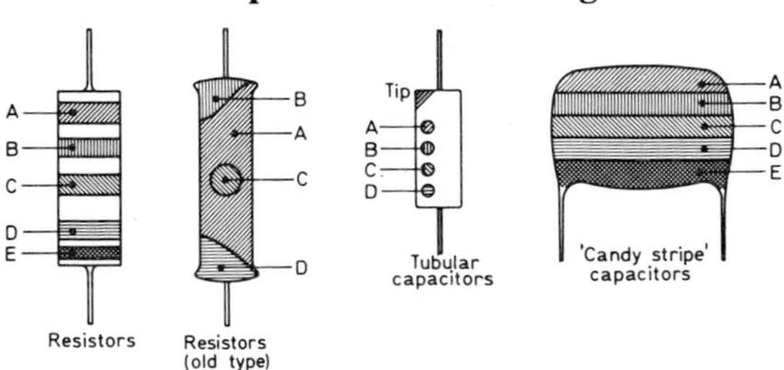

Reactance of capacitors at spot frequencies

	50 Hz	100 Hz	1 kHz	10 kHz	100 kHz	1 MHz	10 MHz	100 MHz
1 pF	—	—	—	—	1·6 M	160 k	16 k	1·6 k
10 pF	—	—	—	1·6 M	160 k	16 k	1·6 k	160
50 pF	—	—	3·2 M	320 k	32 k	3·2 k	320	32
250 pF	—	6·4 M	640 k	64 k	6·4 k	640	64	6·4
1,000 pF	3·2 M	1·6 M	160 k	16 k	1·6 k	160	16	1·6
2,000 pF	1·6 M	800 k	80 k	8 k	800	80	8	0·8
0·01 μF	320 k	160 k	16 k	1·6 k	160	16	1·6	0·16
0·05 μF	64 k	32 k	3·2 k	320	32	3·2	0·32	—
0·1 μF	32 k	16 k	1·6 k	160	16	1·6	0·16	—
1 μF	3·2 k	1·6 k	160	16	1·6	0·16	—	—
2·5 μF	1·3 k	640	64	6·4	0·64	—	—	—
5 μF	640	320	32	3·2	0·32	—	—	—
10 μF	320	160	16	1·6	0·16	—	—	—
30 μF	107	53	5·3	0·53	—	—	—	—
100 μF	32	16	1·6	0·16	—	—	—	—
1,000 μF	3·2	1·6	0·16	—	—	—	—	—

Values above 10 MΩ and below 0·1 Ω not shown. Values in ohms.

Reactance of inductors at spot frequencies

	50 Hz	100 Hz	1 kHz	10 kHz	100 kHz	1 MHz	10 MHz	100 MHz
1 μH	—	—	—	—	0·63	6·3	63	630
5 μH	—	—	—	0·31	3·1	31	310	3·1 k
10 μH	—	—	—	0·63	6·3	63	630	6·3 k
50 μH	—	—	0·31	3·1	31	310	3·1 k	31 k
100 μH	—	—	0·63	6·3	63	630	6·3 k	63 k
250 μH	—	0·16	1·6	16	160	1·6 k	16 k	160 k
1 mH	0·31	0·63	6·3	63	630	6·3 k	63 k	630 k
2·5 mH	0·8	1·6	16	160	1·6 k	16 k	160 k	1·6 M
10 mH	3·1	6·3	63	630	6·3 k	63 k	630 k	6·3 M
25 mH	8	16	160	1·6 k	16 k	160 k	1·6 M	—
100 mH	31	63	630	6·3 k	63 k	630 k	6·3 M	—
1 H	310	630	6·3 k	63 k	630 k	6·3 M	—	—
5 H	1·5 k	3·1 k	31 k	310 k	3·1 M	—	—	—
10 H	3·1 k	6·3 k	63 k	630 k	6·3 M	—	—	—
100 H	31 k	63 k	630 k	6·3 M	—	—	—	—

Values above 10 MΩ and below 0·1 Ω not shown. Values in ohms.

Transistor letter symbols

Bipolar

C_{cb}, C_{ce}, C_{eb} Interterminal capacitance (collector-to-base, collector-to-emitter, emitter-to-base).

C_{ibo}, C_{ieo} Open-circuit input capacitance (common-base, common-emitter).

C_{ibs}, C_{ieo} Short-circuit input capacitance (common-base, common-emitter).

C_{obo}, C_{oeo} Open-circuit output capacitance (common-base, common-emitter).

C_{obs}, C_{oes} Short-circuit output capacitance (common-base, common-emitter).

C_{rbs}, C_{res} Short-circuit reverse transfer capacitance (common-base, common-emitter).

C_{tc}, C_{te} Depletion-layer capacitance (collector, emitter).

f_{hfb}, h_{fe} Small-signal short-circuit forward current transfer ratio cutoff frequency (common-base, common-emitter).

f_{max} Maximum frequency of oscillation.

f_T Transition frequency or frequency at which small-signal forward current transfer ratio (common-emitter) extrapolates to unity.

f_1 Frequency of unity current transfer ratio.

G_{PB}, G_{PE} Large-signal insertion power gain (common-base, common-emitter).

G_{pb}, G_{pe} Small-signal insertion power gain (common-base, common-emitter).

G_{TB}, G_{TE} Large-signal transducer power gain (common-base, common-emitter)

G_{tb}, G_{te} Small-signal transducer power gain (common-base, common-emitter).

h_{FB}, h_{FE} Static forward current transfer ratio (common-base, common-emitter).

h_{fb}, h_{fe} Small-signal short-circuit forward current transfer ratio (common-base, common-emitter).

h_{ib}, h_{ie} Small-signal short-circuit input impedance (common-base, common-emitter).

$h_{ie(imag)}$ or $Im(h_{ie})$ Imaginary part of the small-signal short-circuit input impedance (common-emitter).

$h_{ie(real)}$ or $Re(h_{ie})$ Real part of the small-signal short-circuit input impedance (common-emitter).

h_{ob}, h_{oe} Small-signal open-circuit output admittance (common-base, common-emitter).

$h_{oe(imag)}$ or $Im(h_{oe})$ Imaginary part of the small-signal open-circuit output admittance (common-emitter).

$h_{oe(real)}$ or $Re(h_{oe})$ Real part of the small-signal open-circuit output admittance (common-emitter).

h_{rb}, h_{re} Small-signal open-circuit reverse voltage transfer ratio (common-base, common-emitter).

I_B, I_C, I_E Current, d.c. (base-terminal, collector-terminal, emitter-terminal).

I_b, I_c, I_e Current, r.m.s. value of alternating component (base-terminal, collector-terminal, emitter-terminal).

i_B, i_C, i_E Current, instantaneous total value (base-terminal, collector-terminal, emitter-terminal).

I_{BEV} Base cutoff current, d.c.

I_{CBO} Collector cutoff current, d.c., emitter open.

$I_{E1E2(off)}$ Emitter cutoff current.

I_{EBO} Emitter cutoff current, d.c., collector open.

$I_{Ec(ofs)}$ Emitter-collector offset current.

I_{ECS} Emitter cutoff current, d.c., base-short-circuited to collector.

P_{IB}, P_{IE} Large-signal input power (common-base, common-emitter).

P_{ib}, P_{ie} Small-signal input power (common-base, common-emitter).

P_{OB}, P_{OE} Large-signal output power (common-base, common-emitter).

P_{ob}, P_{oe} Small-signal output power (common-base, common-emitter).

P_T Total nonreactive power input to all terminals.

$r_b'C_c$ Collector-base time constant.

$r_{CE(sat)}$ Saturation resistance, collector-to-emitter.

$Re(y_{ie})$

$Re(y_{oe})$

$r_{e1e2(on)}$ Small-signal emitter-emitter on-state resistance.

R_θ Thermal resistance.

T_j Junction temperature.

t_d Delay time.

t_f Fall time.

t_{off} Turn-off time.

t_{on} Turn-on time.

t_p Pulse time.

t_r Rise time.

t_s Storage time.

t_w Pulse average time.

V_{BB}, V_{CC}, V_{EE} Supply voltage, d.c. (base, collector, emitter).

V_{BC}, V_{BE}, V_{CB}, V_{CE}, V_{EB}, V_{EC} Voltage, d.c. or average (base-to-collector, base-to-emitter, collector-to-base, collector-to-emitter, emitter-to-base, emitter-to-collector).

v_{bc}, v_{be}, v_{cb}, v_{ce}, v_{eb}, v_{ec} Voltage, instantaneous value of alternating component (base-to-collector, base-to-emitter, collector-to-base, collector-to-emitter, emitter-to-base, emitter-to-collector).

$V_{(BR)CBO}$ (formerly BV_{CBO}) Breakdown voltage, collector-to-base, emitter open.

V_{RT} Reach-through (punch-through) voltage.

y_{fb}, y_{fe} Small-signal short-circuit forward-transfer admittance (common-base, common-emitter).

y_{ib}, y_{ie} Small-signal short-circuit input admittance (common-base, common-emitter).

$y_{ie(imag)}$ or $Im(y_{ie})$ Imaginary part of the small-signal short circuit input admittance (common-emitter).

$y_{ie(real)}$ or $Re(y_{ie})$ Real part of the small-signal short-circuit input admittance (common-emitter).

y_{ob}, Y_{oe} Small-signal short-circuit output admittance (common-base, common-emitter).

$y_{oe(imag)}$ or $Im(y_{oe})$ Imaginary part of the small-signal short-circuit output admittance (common-emitter).

$y_{oe(real)}$ or $Re(y_{oe})$ Real part of the small-signal short-circuit output admittance (common-emitter).

y_{rb}, y_{re} Small-signal short-circuit reverse transfer admittance (common-base, common-emitter).

Unijunction

η Intrinsic standoff ratio.

$I_{B_2(mod)}$ Interbase modulated current.

I_{EB_2O} Emitter reverse current.

I_p Peak-point current.

I_V Valley-point current.

r_{BB} Interbase resistance.

T_j Junction temperature.

t_p Pulse time.

t_w Pulse average time.

$V_{B_2B_1}$ Interbase voltage.

$V_{EB_1(sat)}$ Emitter saturation voltage.

V_{OB_1} Base-1 peak voltage.

V_p Peak-point voltage.

V_v Valley-point voltage.

Field Effect

b_{fs}, b_{is}, b_{os}, b_{rs} Common-source small-signal (forward transfer, input, output, reverse transfer) susceptance.

C_{ds} Drain-source capacitance.

c_{du} Drain-substrate capacitance.

C_{iss} Short-circuit input capacitance, common-source.

C_{oss} Short-circuit output capacitance, common-source.

C_{rss} Short-circuit reverse transfer capacitance, common-source.

$\bar{F}$ **or F** Noise figure, average or spot.

g_{fs}, g_{is}, g_{os}, g_{rs} Signal (forward transfer, input, output, reverse transfer) conductance.

G_{pg}, G_{ps} Small-signal insertion power gain (common-gate, common-source).

G_{tg}, G_{ts} Small-signal transducer power gain (common-gate, common-source).

$I_{D(off)}$ Drain cutoff current.

$I_{D(on)}$ On-state drain current.

I_{DSS} Zero-gate-voltage drain current.

I_G Gate current, d.c.

I_{GF} Forward gate current.

I_{GR} Reverse gate current.

I_{GSS} Reverse gate current, drain short-circuited to source.

I_{GSSF} Forward gate current, drain short-circuited to source.

I_{GSSR} Reverse gate current, drain short-circuited to source.

I_n Noise current, equivalent input.

Im(y_{rs}), Im(y_{is}), Im(y_{os}), Im(y_{rs}).

I_s Source current, d.c.

$I_{S(off)}$ Source cutoff current.

I_{SDS} Zero-gate-voltage source current.

$r_{ds(on)}$ Small-signal drain-source on-state resistance.

$r_{DS(on)}$ Static drain-source on-state resistance.

$t_{d(on)}$ Turn-on delay time.

t_f Fall time.

Common transistor and diode data

Bipolar transistors

Type	Case	POL MAT	Vce	Vcb	IC mA	Vces	IC mA	Hfe
AC107	GT3	NG	15	15	10			30-160
AC125	TO-1	PG	12	32	100			100
AC126	TO-1	PG	12	32	100			140
AC127	TO-1	NG	12	32	500			105
AC128	TO-1	PG	16	32	1000	·6	1A	60-175
AC132	TO-1	PG	12	32	200	·35	200	115
AC187	TO-1	NG	15	25	2000	·8	1A	100-500
AC188	TO-1	PG	15	25	2000	·6	1A	100-500
AD149	TO-3	PG	30	50	3500	·7	3A	30-100
AD161	PT1	NG	20	32	3000	·6	1A	80-320
AD162	PT1	PG	20	32	3000	·4	1A	80-320

t_{off} Turn-off time.

t_{on} Turn-on time.

t_p Pulse time.

t_r Rise time.

t_w Pulse average time.

$V_{(BR)GSS}$ Gate-source breakdown voltage.

$V_{(BR)GSSF}$ Forward gate-source breakdown voltage.

$V_{(BR)GSSR}$ Reverse gate-source breakdown voltage.

V_{DD}, V_{GG}, V_{SS} Supply voltage, d.c. (drain, gate, source).

V_{DG} Drain-gate voltage.

V_{DS} Drain-source voltage.

$V_{DS(on)}$ Drain-source on-state voltage.

V_{DU} Drain-substrate voltage.

V_{GS} Gate-source voltage.

V_{GSF} Forward gate-source voltage.

V_{GSR} Reverse gate-source voltage.

$V_{GS(off)}$ Gate-source cutoff voltage.

$V_{GS(th)}$ gate-source threshold voltage.

V_{GU} Gate-substrate voltage.

V_n Noise voltage equivalent input.

V_{SU} Source-substrate voltage.

y_{fs} Common-source small-signal short-circuit forward transfer admittance.

y_{is} Common-source small-signal short-circuit input admittance.

y_{os} Common-source small-signal short-circuit output admittance.

IC mA	Ft MHz	IC mA	Ptot mW	Use	Comparable types
3	2	3	80	Low noise audio	AC125–2N406
2	1·3	10	216	Audio driver	2N406
2	1·7	10	216	Audio driver	2N406
50	1·5	10	340	Audio O/P	AC187
300	1	10	260	Audio O/P	AC188
50	1·3	10	216	Audio O/P	AC188
300	1	10	800	Audio O/P	AC127
300	1	10	220	Audio O/P	AC128
1A	·3	500	32W	GP O/P	OC26,AU106
500	·02	300	4W	Audio amp.	AD165,2N1218,2N1292
500	·015	300	6W	Audio amp.	AD143,AD152,AD427

Type	Case	POL MAT	Vce	Vcb	IC mA	Vces	IC mA	Hfe
AF114	TO-7	PG	15	32	10			150
AF115	TO-7	PG	15	32	10			150
AF116	TO-7	PG	15	32	10			150
AF117	TO-7	PG	15	32	10			150
AF118	TO-7	PG	20	70	30	5	30	35
ASZ15	TO-3	PG	60	100	10A	·4	10A	20-55
ASZ16	TO-3	PG	32	60	10A	·4	10A	45-130
ASZ17	TO-3	PG	32	60	10A	·4	10A	25-75
ASZ18	TO-3	PG	32	100	10A	·4	10A	30-110
BC107	TO-18	NS	45	50	100	·2	100	110-450
BC108	TO-18	NS	20	30	100	·2	100	110-800
BC109	TO-18	NS	20	30	100	·2	100	200-800
BC109C	TO-18	NS	20	30	100	·2	100	420-800
BC157	SOT-25	PS	45	50	100	·25	100	75-260
BC158	SOT-25	PS	25	30	100	·25	100	75-500
BC159	SOT-25	PS	20	25	100	·25	100	125-500
BC177	TO-18	PS	45	50	100	·25	100	75-260
BC178	TO-18	PS	25	30	100	·25	100	75-500
BC179	TO-18	PS	20	25	100	·25	100	125-500
BC182(L)	SOT-30 (TO-92/74)	NS	50	10	200	·25	10	100-480
BC183(L)	SOT-30 (TO-92/74)	NS	30	45	200	·25	10	100-850
BC184(L)	SOT-30 (TO-92/74)	NS	30	45	200	·25	10	250-850
BC186	TO-18	PS	25	40	200	·5	50	40-200
BC207	TO-106	NS	45	50	200	·25	10	110-220
BC208	TO-106	NS	20	25	200	·25	10	110-800
BC209	TO-106	NS	20	25	200	·25	10	200-800
BC212(L)	SOT-30 (TO-92/74)	PS	50	60	200	·25	10	60-300
BC213(L)	SOT-30 (TO-92/74)	PS	30	45	200	·25	10	80-400
BC214(L)	SOT-30 (TO-92/74)	PS	30	45	200	·25	10	80-400
BC327	TO-92	PS	45	—	1000	0·7	500	100-600
BC337	TO-92	NS	45	—	1000	0·7	500	100-600
BC547	SO7-30	NS	45	50	100	·6	100	110-800
BC548	SO7-30	NS	30	30	100	·6	100	110-800
BC549	SO7-30	NS	30	30	100	·6	100	200-800
BC549C	SO7-30	NS	30	30	100	·6	100	420-800
BC635	TO-92(74)	NS	45	45	1A	·5	500	40-250
BC636	TO-92(74)	PS	45	45	1A	·5	500	40-250
BC639	TO-92(74)	NS	80	100	1A	·5	500	40-160
BC640	TO-92(74)	PS	80	100	1A	·5	500	40-160
BCY70	TO-18	PS	40	50	200	·5	50	50
BCY71	TO-18	PS	45	45	200	·5	50	100-600
BCY72	TO-18	PS	25	25	200	·5	50	50
BD137	TO-12G	NS	60	60	1A	·5	500	40-160
BD138	TO-126	PS	60	60	1A	·5	500	40-160
BD139	TO-126	NS	60	100	1A	·5	500	40-160
BD140	TO-126	PS	80	100	1A	·5	500	40-160
BD262	TO-126	PS	60	60	4A	2·5	1·5A	750
BD263	TO-126	NS	60	80	4A	2·5	1·5A	750

IC mA	Ft MHz	IC mA	Ptot mW	Use	Comparable types
1	75	1	75	H.F. amp.	AF144,AF194,2N3127
1	75	1	75	H.F. amp.	AF146,AF185,2N2273
1	75	1	75	H.F. amp.	AF135,AF136,2N3127
1	75	1	75	H.F. amp.	AF136,AF197,2N5354
10	175	10	375	V.H.F. amp.	BFW20
1A	·2	1A	30W	H.C. sw.	OC28
1A	·25	1A	30W	H.C. sw.	OC29,AD138,AD723
1A	·22	1A	30W	H.C. sw.	OC35,AD424
1A	·22	1A	30W	H.C. sw.	OC36
2	300	10	300	S.S. amp.	BC207,BC147,BC182
2	300	10	300	S.S. amp.	BC208,BC148,BC183
2	300	10	300	Low noise s.s. amp.	BC209,BC149,BC184
2	300	10	300	Low noise high gain	BC209C,BC184C,BC149C
2	150	10	300	S.S. amp.	BC177,BC307,BC212
2	150	10	300	S.S. amp.	BC178,BC308,BC213
2	150	10	300	S.S. amp.	BC179,BC309,BC214
2	150	10	300	S.S. amp.	BC157,BC307,BC212
2	150	10	300	S.S. amp.	BC158,BC308,BC213
2	150	10	300	S.S. amp.	BC159,BC309,BC214
2	150	10	300	S.S. amp.	BC107,BC207,BC147
2	150	10	300	S.S. amp.	BC108,BC208,BC148
2	150	10	300	Low noise, high gain	BC109,BC209,BC149
2	50	50	300	G.P. amp.	BC213,BC177,BC158
2	150	10	300	S.S. amp.	BC107,BC182,BC147
2	150	10	300	S.S. amp.	BC108,BC183,BC148
2	150	10	300	Low noise, high gain	BC109,BC184,BC149
2	200	10	300	S.S. amp.	BC307,BC157,BC177
2	200	10	300	S.S. amp.	BC308,BC158,BC178
2	200	10	300	S.S. amp.	
100	100	10	800	O/P	2N3638
100	200	10	800	O/P	2N3642
2	300	10	500	S.S. amp.	BC107,BC207,BC147
2	300	10	500	S.S. amp.	BC108,BC208,BC148
2	300	10	500	Low noise s. sig.	BC109,BC209,BC149
2	300	10	500	Low noise, high gain	BC109C,BC149C
150	130	500	1W	Audio O/P	BC639
150	130	500	1W	Audio O/P	BC640
150	130		1W	Audio O/P	MU9610,TT801
150	130		1W	Audio O/P	MU9660,TT800
10	250	50	350	G.P.	BC212
10	200	50	350	G.P.	BC212
10	200	50	350	G.P.	BC213
150	250	500	8W	G.P. O/P	BD139
150	75	500	8W	G.P. O/P	BD140
150	250	500	8W	G.P. O/P	40409
150	75	500	8W	G.P. O/P	40410
1·5A	7	1·5A	36W	High gain darl. O/P	BD266
1·5A	7	1·5A	36W	High gain darl. O/P	BD267

Type	Case	POL MAT	Vce	Vcb	IC mA	Vces	IC mA	Hfe
BD266A	TO-220	PS	80	80	8A	2	3A	750
BD267A	TO-220	NS	80	100	8A	2	3A	750
BDX64A	TO-3	PS	80	80	12A	2·5	5A	1000
BDX65A	TO-3	NS	80	80	12A	2·5	5A	1000
BDY20	TO-3	NS	60	100	15A	1·1	4A	20-70
BF115	TO-72(28)	NS	30	50	30			45-165
BF167	TO-72(28)	NS	30	40	25			26
BF173	TO-72(28)	NS	25	40	25			37
BF177	TO-39	NS	60	100	50			20
BF178	TO-39	NS	115	185	50			20
BF179	TO-39	NS	115	250	50			20
BF180	TO-72(25)	NS	20	30	20			13
BF184	TO-72(28)	NS	20	30	30			75-750
BF185	TO-72(28)	NS	20	30	30			34-140
BF194	SOT 25/1	NS	20	30	30			65-220
BF195	SOT 25/1	NS	20	30	30			35-125
BF200	TO-72(25)	NS	20	30	20			15
BF336	TO-39	NS	180	185	100			20-60
BF337	TO-39	NS	200	300	100			20-60
BF338	TO-39	NS	225	250	100			20-60
BFY50	TO-39	NS	35	80	1A	2	150	30
BFY51	TO-39	NS	30	60	1A	·35	150	40
BFY52	TO-39	NS	20	40	1A	·35	150	60
MJ2501	TO-3	PS	80	80	10A	2	5A	1000
MJ2955	TO-3	PS	60	70	15A	1·1	4A	20-70
MJ3001	TO-3	NS	80	80	10A	2	5A	1000
MJE2955	90·05	PS	60	70	10A	1·1	4A	20-70
MJE3055	90·05	NS	60	70	10A	1·1	4A	20-70
MU9610	152	NS	30	40	2A	0·4	1·5A	80-400
MU9611	152·01	NS	30	40	2A	0·4	1·5A	80-400
MU9660	152	PS	30	40	2A	0·4	1·5A	80-400
MU9661	152·01	PS	30	40	2A	0·4	1·5A	80-400
NSD106	TO-202(35)	NS	100	140		2·9	100	50-150
NSD206	TO-202(35)	PS	100	100		2·1	100	50-150
OC26	TO-3	PG	30	50	3·5A	·7	3A	30-100
OC28	TO-3	PG	60	100	10A	·4	10A	20-55
OC44N	TO-1	PG	5	15	10			45-225
OC45	GT-3	PG	5	15	10			25-125
OC70	GT-3	PG	10	30	50			30
OC71	GT-3	PG	10	30	50			30-75
OC72	GT-6	PG	16	32	250			45-120
OC74N	TO-1	PG	10	20	300	6	300	60-150
OC75	GT-3	PG	10	30	50			60-130
TIP31B	TOP-66	NS	80	80	3A	1·2	3A	20
TIP32B	TOP-66	PS	80	80	3A	1·2	3A	20
TIP2955	TOP-3	PS	70	100	15A	1·1	4A	20
TIP3055	TOP-3	NS	70	100	15A	1·1	4A	20
2N301	TO-3	PG	32	40	3A			50
2N706A	TO-18	NS	15	25	200			20
2N2926	TO-92(74)	NS	25	25	100			150
2N3053	TO-39	NS	40	60	700	1·4	150	50-250
2N3054	TO-66	NS	55	90	4A	1	200	25
2N3055	TO-3	NS	60	90	15A	1·1	4A	20
2N3563	TO-106	NS	12	30	50			20-200

IC mA	Ft MHz	IC mA	Ptot mW	Use	Comparable types
3A	7		60W	High gain darl. O/P	
3A	7		60W	High gain darl. O/P	
8A	7	5A	117W	Darl. O/P	
8A	7	5A	117W	Darl. O/P	
4A	1	4A	115	Power O/P	2N3055
1	230	1	145	V.H.F. amp.	
4	350	4	130	T.V. I.F. amp.	
7	550	5	230	T.V. I.F. amp.	
15	120	10	795	T.V. video amp.	BF336
30	120	10	1·7W	T.V. vidoo amp.	BF336
20	120	10	1·7W	T.V. video amp.	BF338
2	675	2	150	U.H.F. amp.	BF200
1	300	1	145	H.F. amp.	
1	220	1	145	H.F. amp.	BF195
1	260	1	250	H.F. amp.	
1	200	1	250	H.F. amp.	BF185
3	650	3	150	V.H.F. amp.	BF180
30	130		3W	Video amp.	
30	130		3W	Video amp.	
30	130		3W	Video amp.	
150	60	50	2·86W	G.P.	
150	50	50	2·86W	G.P.	
150	50	50	2·86W	G.P.	
5A			150W	Darl. O/P	
4A	4	500	115W	High power O/P	2N4908,2N4909,2N5871
5A			150W	Darl. O/P	
4A	2	500	90W	High power O/P	TIP2955
4A	2	500	90W	High power O/P	TIP3055
350	70	250	1W	O/P	TT801
350	70	250	1W	O/P	TT801
350	70	250	1W	O/P	TT800
350	70	250	1W	O/P	TT800
100	80	50		Driver–O/P	
100	150	50		Driver–O/P	
1A	3	500	32W	G.P. O/P	AD149
1A	2	1A	30W	H.C. switch	ASZ15
1	7·5	1	85	R.F. amp.	AF125,AF135,AF172
1	3	3	85	R.F. amp.	AF132,AF185,AF196
5	5		125	G.P. amp.	AC121,AC126,2N1190
3	6		125	G.P. amp.	AC126,2N2429
10	35		165	Audio O/P	AC122,AC125,AC162
50	1		550	Audio O/P	AC125,AC180,AC192
3	1		125	G.P. amp.	AC173,AC192
500	3	500	40W	Power amp–Sw	
500	3	500	40W	Power amp–Sw	
4A	8		90W	Power amp–Sw	MJE2955
4A	8		90W	Power amp–Sw	MJE3055
1A	2	1A	11W	Audio O/P	AT1138,OC26
10	200		300	High speed Sw	
2	100		200	G.P.	BC108 etc.
150	100	50	2·86W	G.P. switch	BD137
500	8	200	25W	Audio O/P	TIP31B
4A	8	1A	115W	O/P–Sw	BDY20
8	600	8	200	RF–IF amp	BF173

Type	Case	POL MAT	Vce	Vcb	IC mA	Vces	IC mA	Hfe
2N3564	TO-106	NS	15	30	100	3	20	20-500
2N3565	TO-106	NS	25	30	50	35	1	150-600
2N3566	TO-105	NS	30	40	200	1	100	150-600
2N3567	TO-105	NS	40	80	500	25	150	40-120
2N3568	TO-105	NS	60	80	500	25	150	40-120
2N3569	TO-105	NS	40	80	500	25	150	100-300
2N3638	TO-105	PS	25	25	500	25	50	30
2N3638A	TO-105	PS	25	25	500	25	50	100
2N3640	TO-106	PS	12	12	80	2	10	30-120
2N3641	TO-105	NS	30	60	500	22	150	40-120
2N3642	TO-105	NS	45	60	500	22	150	40-120
2N3643	TO-105	NS	30	60	500	22	150	100-300
2N3644	TO-105	PS	45	45	500	1	300	115-300
2N3645	TO-105	PS	60	60	500	1	300	115-300
2N3702	TO-92(74)	PS	25	40	200	25	50	60-300
2N3904	TO-92	NS	40	60	200			100-300
2N4250	TO-106	PS	40	40	100	25	10	250-400
2N4258	TO-106	PS	12	12	50	5	50	30-120
2N4292	TO-92	NS	15	30	50	6	10	20
2N4403	TO-92	PS	40	40	600			100-300
2N5589	MT-71C	NS	18	36	600			5
2N5590	MT-72C	NS	18	36	2A			5
2N5591	MT-72C	NS	18	36	4A			5
2N5871	TO-3	PS	60	60	7A	1	4A	20-100
40250	TO-66	NS	50	50	4A	1·5	1·5A	25
40408	TO-5	NS	80		700	1·4	150	40-200
40409	TO-39(H)	NS	80		700	1·4	150	50-250
40410	TO-39(H)	PS	80		700	1·4	150	50-250

Common transistor and diode data

FETS

		BV_{GSS}		$V_{GS(OFF)}$				$I_{DSS(mA)}$			
Type	Case	V	$I_{G(uA)}$	Min	Max	V_{DS}	$I_{D(nA)}$	Min	Max	V_{DS}	V_{GS}
MPF102	TO-92(72)	25	10	5	8	15	2	2	20	15	0
MPF103	TO-92(72)	25	1		6	15	1	1	5	15	0
MPF104	TO-92(72)	25	1		7	15	1	2	9	15	0
MPF105	TO-92(72)	25	1		8	15	10	4	16	15	0
MPF106	TO-92(72)	25	1	5	4	15	10	4	10	15	0
2N5457	TO-92(72)	25	1	5	6	15	10	1	5	15	0
2N5458	TO-92(72)	25	1	1	7	15	10	2	6	15	0
2N5459	TO-92(72)	25	1	2	8	15	10	4	9	15	0
2N5484	TO-92(72)	25	1	3	3	15	10	1	5	15	0
2N5485	TO-92(72)	25	1	5	4	15	10	4	10	15	0
BFW10	TO-72(25)	30						8	20	15	0
BFW11	TO-72(25)	30						4	10	15	0
BFW61	TO-72(25)	25						2	20	15	0
MPF121	206	7·20			4	15		5	30	15	0
2N4342	TO-106	20			5	10		12	30	10	0

IC mA	Ft MHz	IC mA	Ptot mW	Use	Comparable types
15	400	15	200	RF–IF amp	BF167
1	40	1	200	Low level amp	BC108,BC208
10	40	30	300	GP amp & Sw	BC183
1	60	50	300	GP amp & Sw	BC337
1	60	50	300	GP amp & Sw	
1	60	50	300	GP amp & Sw	
50	100	50	300	GP amp & Sw	BC327
50	150	50	300	GP amp & Sw	BC558
10	300	10	200	Saturated switch	
	250	50	350	GP amp & Sw	BC337
150	250	50	350	GP amp & Sw	BC337
150	250	50	350	GP amp & Sw	BC337
50	200	20	300	GP amp & Sw	BC327
500	200	20	300	GP amp & Sw	
50	100	50	360	GP amp & Sw	BC213
1mA			310	Low level amp.	BC167A,BF194
1	50		200	Low level amp.	BC559
10	700	10	200	Saturated Sw	
3	600	4	200	Saturated Sw	
10			310	G.P.	BC307A,2N2904
100	175→	3W	15W	H.F. mobile R.F.	
250	175→	10W	30W	H.F. mobile R.F.	
500	175→	25W	70W	H.F. mobile R.F.	
2·5A	4	250	100W	Power transistor	2N5872,2N4908,MJ2955
100	1		29W	Power transistor	2N3054
200	100		1W	Power transistor	BC639
150	100		3W	Power transistor	BD139
150	100		3W	Power transistor	BD140

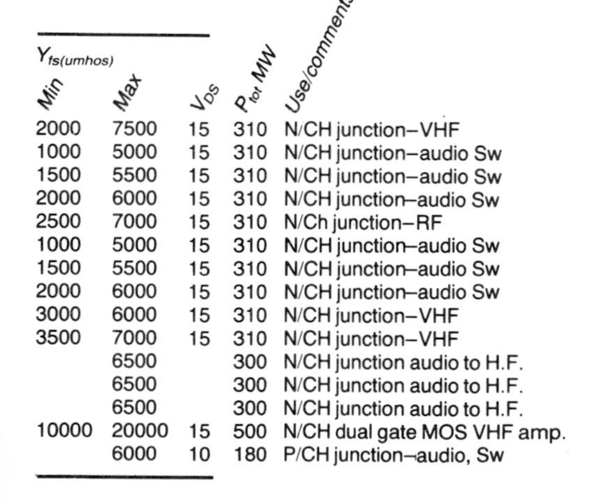

Y_{fs}(umhos) Min	Max	V_{DS}	P_{tot} MW	Use/comments
2000	7500	15	310	N/CH junction–VHF
1000	5000	15	310	N/CH junction–audio Sw
1500	5500	15	310	N/CH junction–audio Sw
2000	6000	15	310	N/CH junction–audio Sw
2500	7000	15	310	N/Ch junction–RF
1000	5000	15	310	N/CH junction–audio Sw
1500	5500	15	310	N/CH junction–audio Sw
2000	6000	15	310	N/CH junction–audio Sw
3000	6000	15	310	N/CH junction–VHF
3500	7000	15	310	N/CH junction–VHF
	6500		300	N/CH junction audio to H.F.
	6500		300	N/CH junction audio to H.F.
	6500		300	N/CH junction audio to H.F.
10000	20000	15	500	N/CH dual gate MOS VHF amp.
	6000	10	180	P/CH junction–audio, Sw

Common transistor and diode data

Power MOSFETS/DMOS and VMOS

Type	Case	Channel	P_{tot}	V_{ds}	V_{dg}	$V_{gs(th)}$
IRF120	TO-3(F)	n	40W	100V	100V	4V
IRF130	TO-3(F)	n	75W	100V	100V	4V
IRF9130	TO-3(F)	p	75W	−100V	−100V	−4V
IRF510	TO220(F)	n	20W	100V	100V	4V
IRF530	TO220(F)	n	75W	100V	100V	4V
IRF640	TO220(F)	n	125W	200V	200V	4V
IRF9520	TO220(F)	p	40W	−100V	−100V	−4V
IRF9530	TO220(F)	p	75W	−100V	−100V	−4V
VN10KM	TO92(F)	n	1W	60V	60V	2·5V max
VN1010	TO92(F)	n	1W	100V	100V	2V max
VN46AF	TO202(F)	n	12·5W	40V	40V	2V max
VN66AF	TO202(F)	n	12·5W	60V	60V	2V max
VN88AF	TO202(F)	n	12·5W	80V	80V	2V max
2SJ50	TO3(F)	p	100W	−160V	−160V	−1·5 max
2SK133	TO3(F)	n	100W	120V	120V	1·5 max

Common transistor and diode data

Unijunction transistors

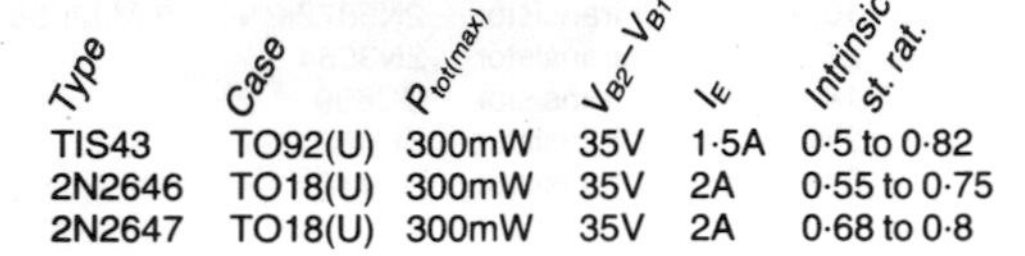

Type	Case	$P_{tot(max)}$	$V_{B2}-V_{B1}$	I_E	Intrinsic st. rat.
TIS43	TO92(U)	300mW	35V	1·5A	0·5 to 0·82
2N2646	TO18(U)	300mW	35V	2A	0·55 to 0·75
2N2647	TO18(U)	300mW	35V	2A	0·68 to 0·8

Common transistor and diode data

Thyristors

Type	Case	PIV	$I_{T(rms)}$ max	V_{GT}	I_{GT}
C106D	TO202(TH)	400V	4A	0·8V	0·2mA
2N3525	TO66(TH)	400V	5A	2V	15mA
2N4443	TO126(TH)	400V	5A	1·5V	30mA
BTX18-400	TO5(TH)	500 V	1A	2V	5mA

Common transistor and diode data

Triacs

Type	Case	PIV	$I_{T(rms)}$	V_{GT}	I_{GT}
TRI400-0·35	TO92(TRI)	400V	0·35A	2V	5mA
C206D	TO202(TRI)	400V	3A	2V	5mA
C226D	TO202(TRI)	400V	8A	2·5V	50mA
C246D	TO202(TRI)	400V	15A	2·5V	50mA

I_{gss}	I_{dss}	gfs(mS)	$I_{d(max)}$
100nA	1mA	1500	6A
100nA	1mA	3000	12A
−100nA	−1mA	2000	−8A
500nA	0·5mA	1000	3A
500nA	1mA	3000	10A
500nA	1mA	6000	11A
−500nA	−1mA	900	−4A
−500nA	−1mA	2000	−7A
10μA	10μA	200	0·5A
10μA	10μA	200	0·5A
10μA	10μA	250	2A
10μA	10μA	250	2A
10μA	10μA	250	2A
−10μA	−10μA	1000	−7A
10μA	10μA	1000	7A

Thyristors

Some confusion exists over the use of the term *thyristor*. Here it is taken as a generic title, the *silicon controlled rectifier* (s.c.r.) being one particular type. An s.c.r. with its equivalent circuit and symbol is shown below. With G open-circuit Q_1 is cut off, so the device will not conduct unless V_{AK} is made sufficiently positive for it to avalanche. If the gate is made positive I_{C1} flows. Now I_{C1} is base current for Q_2 and I_{C2} is base current for Q_1. Therefore I_{C1} brings on Q_2 and the cumulative action makes both transistors saturate, i.e. the voltage across the device falls to a minimum and it is in the condition of a closed switch. Once the s.c.r. is on, the gate voltage is not required to sustain it. In fact the gate cannot be used to turn it off, and this must be done by reducing V_{AK} to a very small voltage.

To switch a thyristor on without any gate voltage the device must be made to avalanche. Such a device, which will avalanche in either direction in a controlled manner, is the *bi-directional diode* thyristor or *diac*. A gated version of this is the *bi-directional triode* thyristor or *triac*, which will conduct in either direction when gated with a pulse of the appropriate polarity. In this way it performs as a pair of thyristors connected in inverse parallel.

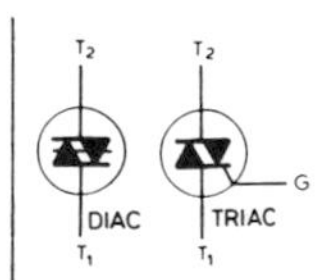

T_1 = MAIN TERMINAL 1
T_2 = MAIN TERMINAL 2
G = GATE

SILICON CONTROLLED RECTIFIER WITH EQUIVALENT CIRCUIT AND SYMBOL

Common transistor and diode data

Rectifiers/diodes

Type	Mat	V_R	$I_F(A)$	V_F	$I_F(A)$	$I_R(\mu A)$	V_R
A14P	S	1000	2·5	1·25	2·5	0·5	1000
A15A	S	100	5	1·1	5	5	100
BYX21L/200R	S	75	25	1·2	25	1·1	75
EM4005	S	50	1	1·1	1	5	50
EM401	S	100	1	1·1	1	5	100
EM404	S	400	1	1·1	1	5	400
EM410	S	1000	1	1·1	1	5	1000
1N4001	S	50	1	1·1	1	5	50
1N4002	S	100	1	1·1	1	5	100
1N4004	S	400	1	1·1	1	5	400
1N4007	S	1000	1	1·1	1	5	1000
1N5408	S	1000	3	1	3	5	1000
1N5059(A14B)	S	200	2·5	1·25	2·5	0·2	200
1N5060(A14D)	S	400	2·5	1·25	2·5	0·2	400
1N5061(A14M)	S	600	2·5	1·25	2·5	0·2	600
1N5062(A14N)	S	800	2·5	1·25	2·5	0·2	800
MR110	S	100	10				
MR410	S	400	10				

Common transistor and diode data

Diodes

Type	Case	V_R	$I_F(mA)$	$C_d(pF)$	V_F	$I_F(mA)$
Germanium						
AA119	DO-7	30	100	1·2	2·2	10
OA90	DO-7	20	45		1·5	10
OA91	DO-7	90	150		1·9	10
OA95	DO-7	90	150		1·5	10
Silicon						
BA100	DO-7	60	90	25	96	10
BA102	DO-7	20		20-45	C_d ratio 1·4 @ 4/10 V/VV	
BA114	DO-7		20		7	1
OA200	DO-7	50	160	25	96	10
OA202	DO-7	150	160	25	96	10
1N914A	DO-35	75	75	4	1	10
1N4148	SD-5	75	75	4	1	10
50822800	DO-7	70	15	2	41	1

Use
Transient protected (controlled avalanche)
G.P. rectifier
Automobile H. duty
G.P. rectifier
G.P. rectifier
G.P. rectifier
G.P. rectifier
G.P. rectifier
G.P. rectifier
G.P. rectifier
G.P. rectifier
G.P. rectifier
Transient protected (controlled avalanche)
Transient protected (controlled avalanche)
Transient protected (controlled avalanche)
Transient protected (controlled avalanche)
G.P. stud mount
G.P. stud mount

$I_R(\mu A)$	V_R	$T_{rr}(nS)$	Use	Comparable types
150	30		AM/FM detection	
			Point contact	
450	20		G.P.–point contact	OA70,OA80
180	75		G.P.–point contact	OA71,OA79,OA81
110	75		G.P.–point contact	
10	60		G.P.–alloyed	
			Variable capacitance	
			Bias stabilizer	
1	50		Small signal–alloyed	
1	150		Small signal–alloyed	
5	75	4	Small signal–switching	1N4148
·025	20	4	Small signal–switching	1N014A
0·2	50	0·1	Schottky (hot carrier)	
			UHF detector, mixer, switch	

Bridge rectifier data

Type	Case	PIV	$V_{IN(rms)max}$	$V_{F(max)}$	$I_{F(ave)}$
Vm28	1	200V	140V	1·9V at 1A	0·9A
Vm48	1	400V	280V	1·9V at 1A	0·9A
Vm88	1	800V	560V	1·9V at 1A	0·9A
Wo05	2	50V	35V	2V at 1A	1A
Wo2	2	200V	140V	2V at 1A	1A
Wo4	2	400V	280V	2V at 1A	1A
Wo8	2	800V	560V	2V at 1A	1A
BY164	3	60V	42V	2V at 1A	1A
So05	4	50V	35V	2V at 1A	2A
So4	4	400V	280V	2V at 1A	2A
SKB2/02L5A	3	200V	140V	1·8V at 1A	1·6A
SKB2/04L5A	3	400V	280V	1·8V at 1A	1·6A
KBLo2	5	200V	140V	1·2V at 1A	3A
KBLo8	5	800V	560V	1·2V at 1A	3A
Ko1	6	100V	70V	2·1V at 10A	25A
Ko2	6	400V	280V	2·1V at 10A	25A
KBPC3502	6	200V	140V	1·2V at 17·5A	35A
KBPC3506	6	600V	420V	1·2V at 17·5A	35A

Voltage regulator data

Type	Case	$I_{out(max)}$	V_{out}	$V_{in(range)}$	Load reg.
78L05	2A	100mA	5V	7 to 25V	0·2%
79L05	2B	−100mA	−5V	7 to 25V	0·2%
78L12	2A	100mA	12V	14·5 to 35V	0·2%
79L12	2B	−100mA	−12V	14·5 to 35V	0·2%
78L15	2A	100mA	15V	17·5 to 35V	0·3%
79L15	2B	−100mA	−15V	17·5 to 35V	0·3%
78L24	2A	100mA	24V	27 to 35V	0·4%
79L24	2B	−100mA	−24V	27 to 35V	0·4%
7805	1A	1A	5V	7 to 25V	0·2%
7905	1B	−1A	−5V	7 to 25V	0·2%
7812	1A	1A	12V	14·5 to 30V	0·4%
7912	1B	−1A	−12V	14·5 to 30V	0·4%
7815	1A	1A	15V	17·5 to 30V	0·5%
7915	1B	1A	−15V	17·5 to 30V	0·5%
7824	1A	1A	24V	27 to 38V	0·6%
7924	1B	1A	−24V	27 to 38V	0·6%
LM309K	4A	1·2A	5V	7 to 35V	1%
78H05	4B	5A	5V	8 to 25V	0·2%
78H12	4B	5A	12V	15 to 25V	0·2%
78HG	5A	5A	5 to 24V	8 to 40V	1%
79HG	5B	5A	−2 to −24V	8 to 40V	0·7%
317K	4B	1·5A	1·2 to 37V	4 to 40V	0·1%
338K	4B	5A	1·2 to 32V	4 to 35V	0·1%
396K	4C	10A	1·25 to 15V	4 to 25V	0·15%
4195NB	6	±50mA	±15V	±18 to 30V	0·6%
78MGU1C	3A	500mA	5V to 30V	8 to 40V	1%
79MGU1C	3B	−500mA	−3 to −30V	7 to 30V	1%

Bridge rectifier encapsulations

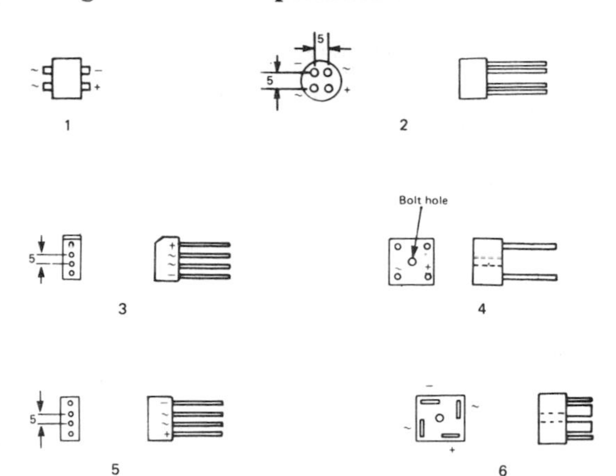

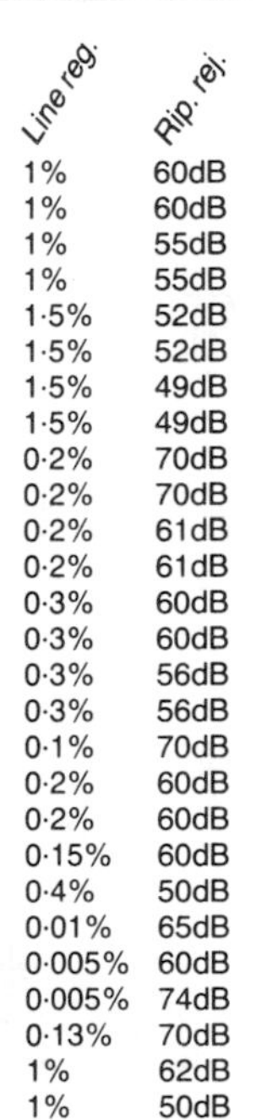

Line reg.	Rip. rej.
1%	60dB
1%	60dB
1%	55dB
1%	55dB
1·5%	52dB
1·5%	52dB
1·5%	49dB
1·5%	49dB
0·2%	70dB
0·2%	70dB
0·2%	61dB
0·2%	61dB
0·3%	60dB
0·3%	60dB
0·3%	56dB
0·3%	56dB
0·1%	70dB
0·2%	60dB
0·2%	60dB
0·15%	60dB
0·4%	50dB
0·01%	65dB
0·005%	60dB
0·005%	74dB
0·13%	70dB
1%	62dB
1%	50dB

Voltage regulator encapsulations

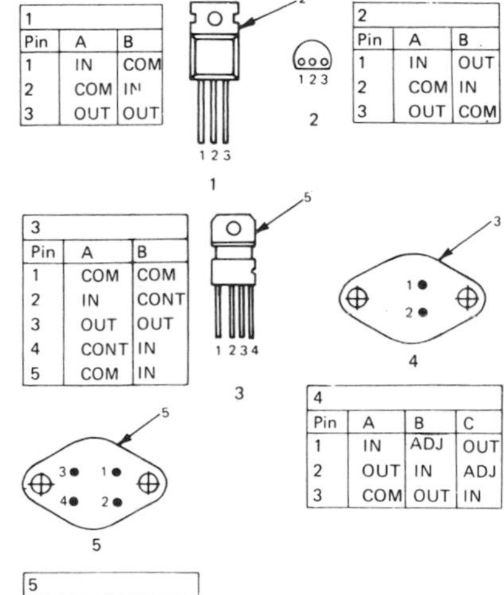

1		
Pin	A	B
1	IN	COM
2	COM	IN
3	OUT	OUT

2		
Pin	A	B
1	IN	OUT
2	COM	IN
3	OUT	COM

3		
Pin	A	B
1	COM	COM
2	IN	CONT
3	OUT	OUT
4	CONT	IN
5	COM	IN

4			
Pin	A	B	C
1	IN	ADJ	OUT
2	OUT	IN	ADJ
3	COM	OUT	IN

5		
Pin	A	B
1	OUT	CONT
2	CONT	OUT
3	IN	COM
4	COM	IN
5	ISOL	ISOL

Zener diodes

BZY88C series
Tolerance: ±5% Maximum dissipation: 500 mW
Range values: 2·7V; 3V; 3·3V; 3·6V; 3·9V;4·3V; 4·7V; 5·1V; 5·6V; 6·2V; 6·8V; 7·5V; 8·2V; 9·1V; 10V; 11V; 12V; 13V; 15V; 16V; 18V; 20V; 22V; 24V; 27V; 30V

BZX 85 series
Tolerance: ±5% Maximum dissipation: 1·3W
Range values: 2·7V; 3·0V; 3·3V; 3·6V; 3·9V; 4·3V; 4·7V; 5·1V; 5·6V; 6·2V; 6·8V

BZX 61 series
Tolerance: ±5% Maximum dissipation: 1·3W
Range values: 4·7V; 5·1V; 5·6V; 6·2V; 6·8V; 7·5V; 8·2V; 9·1V; 10V; 11V; 12V; 13V; 15V; 16V; 18V; 20V; 22V; 24V; 27V; 30V; 33V; 36V; 39V; 43V; 47V; 51V; 56V; 62V; 68V; 75V

1N5333 series
Tolerance: ±5% Maximum dissipation: 5W
Range values: 3·3V; 3·9V; 4·7V; 5·6V; 6·8V; 8·2V; 9·1V; 10V; 12V; 15V; 24V

Transistor and diode encapsulations

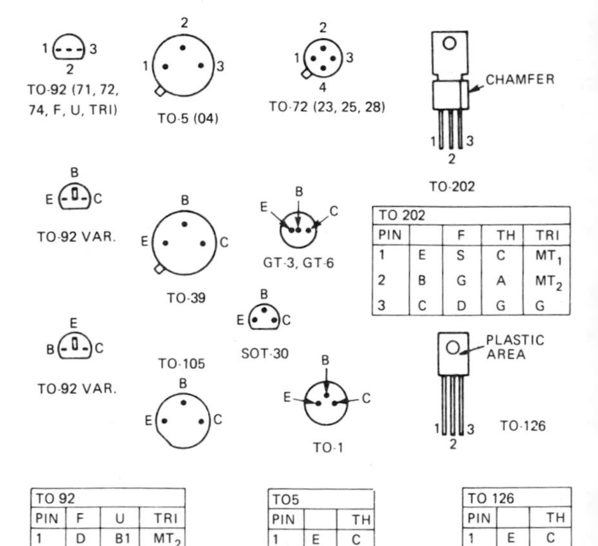

TO 202				
PIN		F	TH	TRI
1	E	S	C	MT_1
2	B	G	A	MT_2
3	C	D	G	G

TO 92			
PIN	F	U	TRI
1	D	B1	MT_2
2	G	B2	G
3	S	E	MT_1

TO5		
PIN		TH
1	E	C
2	B	G
3	C	A

TO 126		
PIN		TH
1	E	C
2	C	A
3	B	G

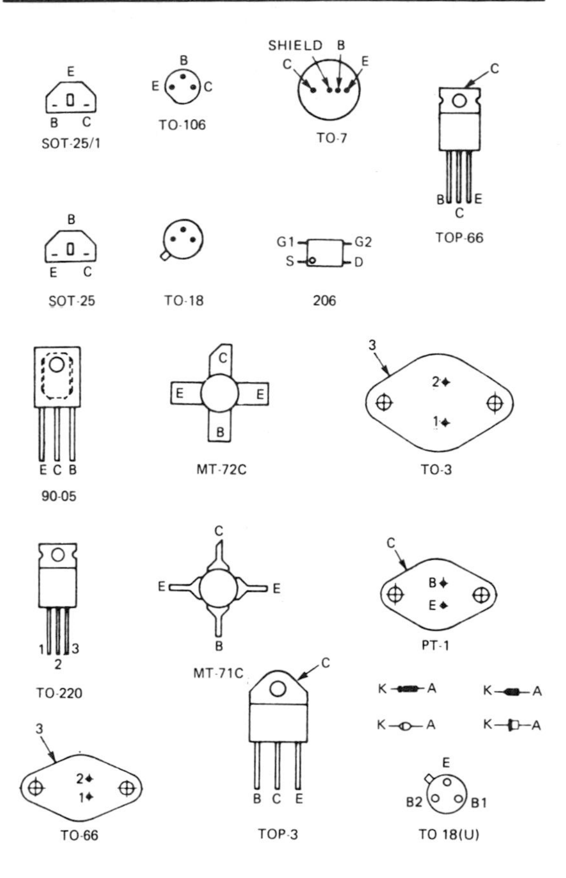

PIN	71	
	T	FET
1	C	G
2	E	D
3	B	S

PIN	72 (STD)	
	T	FET
1	C	G
2	B	S
3	E	D

PIN	74	
	T	FET
1	B	S
2	C	G
3	E	D

PIN	T (25)	FET N (25)
1	E	S
2	B	D
3	C	G
4	GND	CASE

PIN	T (28)	FET P (23)
1	B	S
2	E	G
3	C	D
4	GND	CASE

TO 3		
Pin		F
1	E	G
2	B	S
3	C	D

TO 66		
Pin		TH
1	E	C
2	B	G
3	C	A

TO 220		
Pin		F
1	B	G
2	C	D
3	E	S

BIPOLARS

E : EMITTER
B : BASE
C : COLLECTOR
NG : NPN GERMANIUM
PG : PNP GERMANIUM
NS : NPN SILICON
PS : PNP SILICON

FETS

S : SOURCE
G : GATE
D : DRAIN
N/CH : N CHANNEL
P/CH : P CHANNEL

DIODES

A : ANODE
K : CATHODE

G P : GENERAL PURPOSE
S S : SMALL SIGNAL
SW : SWITCH
O/P : OUTPUT
R F : RADIO FREQUENCY
H F : HIGH FREQUENCY
V H F : VERY HIGH FREQUENCY

Pro Electron system of semiconductor type labelling

The Pro Electron system of semiconductor labelling, used by most European manufacturers, describes a device by means of a code comprising two letters followed by a serial number. The letters define the semiconductor material used and the device's general function, as listed below:

First letter	Semiconductor material
A	Germanium
B	Silicon
C	Gallium arsenide or similar
D	Indium antimonide or similar
R	Cadmium sulphide or similar

Second letter	General function
A	Detection diode, high speed, diode, mixer diode
B	Varicap diode
C	Audio frequency, non-power, transistor
D	Audio frequency power transistor
E	Tunnel diode
F	Radio frequency, non-power, transistor
G	Miscellaneous
L	Radio frequency power transistor
N	Photo-coupler
P	Radiation (e.g., light) detecting device
Q	Radiation source
R	Switching device, non-power
S	Switching transistor, non-power
T	Switching device, power
U	Switching transistor, power
X	Multiplier diode
Y	Rectifying diode or similar device
Z	Voltage reference or regulating diode

The serial number defines the device's particular application, and will consist of either: three numbers (which shows the device is intended for use primarily in consumer applications), or; a letter followed by two numbers (which shows the device is intended for use primarily in industrial or professional environments).

Range numbers

Where variants of a device exist, the above code is addended with a further code (separated by a hyphen) to identify the specific device type within the range. Two classes of device are affected:

(a) Rectifer diodes and thyristors; the group of figure indicate either the repetitive peak inverse voltage, V_{RRM}, or the repetitive peak off-state voltage, V_{DRM}, whichever is the lowest.

(b) Voltage regulator diodes and transient suppression diodes; a first letter (voltage regulator diodes only) indicates operating voltage tolerance, where: A = ±1%

B = ±2%
C = ±5%
D = ±10%
E = ±15%

and a group of figures indicate the typical operating voltage (or the maximum recommended stand-off voltage, in the case of transient suppressor diodes).

In all cases, a final letter (R) may be used, to indicate a reverse polarity version (i.e., one with a stud anode).

Component symbols (BS 3939)

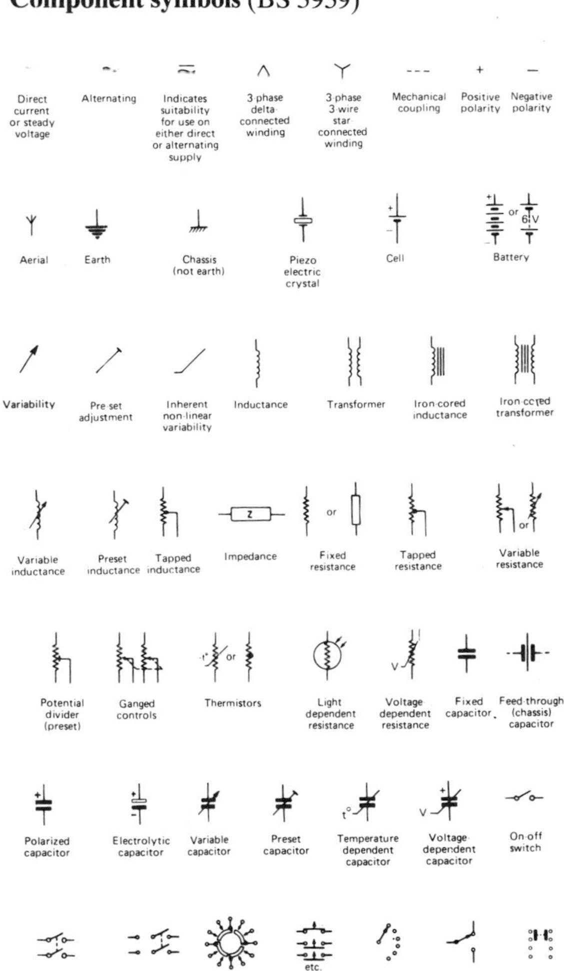

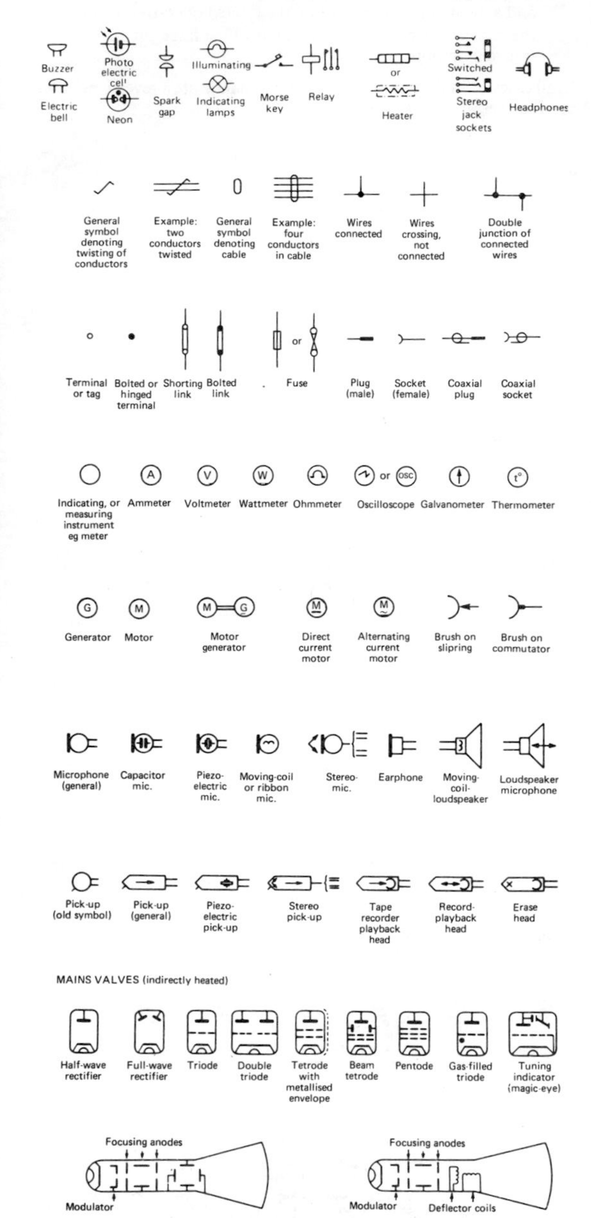
Buzzer
Electric bell
Photo electric cell
Neon
Spark gap
Illuminating
Indicating lamps
Morse key
Relay
or
Heater
Switched
Stereo jack sockets
Headphones
General symbol denoting twisting of conductors
Example: two conductors twisted
General symbol denoting cable
Example: four conductors in cable
Wires connected
Wires crossing, not connected
Double junction of connected wires
Terminal or tag
Bolted or hinged terminal
Shorting link
Bolted link
or
Fuse
Plug (male)
Socket (female)
Coaxial plug
Coaxial socket
A
V
W
or
OSC
t°
Indicating, or measuring instrument eg meter
Ammeter
Voltmeter
Wattmeter
Ohmmeter
Oscilloscope
Galvanometer
Thermometer
G
M
M
G
M
M
Generator
Motor
Motor generator
Direct current motor
Alternating current motor
Brush on slipring
Brush on commutator
Microphone (general)
Capacitor mic.
Piezo-electric mic.
Moving-coil or ribbon mic.
Stereo-mic.
Earphone
Moving-coil-loudspeaker
Loudspeaker microphone
Pick-up (old symbol)
Pick-up (general)
Piezo-electric pick-up
Stereo pick-up
Tape recorder playback head
Record-playback head
Erase head
MAINS VALVES (indirectly heated)
Half-wave rectifier
Full-wave rectifier
Triode
Double triode
Tetrode with metallised envelope
Beam tetrode
Pentode
Gas-filled triode
Tuning indicator (magic-eye)
Focusing anodes
Modulator
Electrostatic cathode-ray tube
Focusing anodes
Modulator
Deflector coils
Electromagnetic cathode-ray tube

SEMICONDUCTORS: may be shown with envelope

e.g.

or without envelope

e.g.

Separate envelopes are not usually shown for semiconductors forming part of an integrated circuit. Semiconductor junctions may be shown in solid form

e.g.

or in outline

e.g.

pn Diode | Zener diode | Tunnel diode | Bidirectional diode (diac) | Temperature dependent diode | Capacitive diode (varactor) | Controlled rectifier p-gate (thyristor)

Controlled rectifier n-gate (thyristor) | Triac | PNP transistor | NPN transistor | NPN transistor with collector connected to case | Unijunction transistor n type base | Unijunction transistor p type base

Light-sensitive diode | Photo-conductive cell | Light sensitive transistor | Light emitting diode | Op amp | n-channel junction gate (JFET) | p-channel JFET

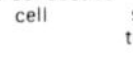

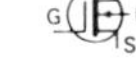

Three terminal depletion-type IGFET, substrate tied to source | Three terminal p-channel IGFET | Four terminal depletion-type IGFET | Four terminal enhancement-type IGFET | Five terminal dual-gate depletion-type IGFET | Five terminal dual-gate enhancement-type IGFET

SOUND ELECTRONIC DEVICES

Recording or reproducing; arrow points in direction of energy transfer | Recording and reproducing, radiating and receiving | Magneto-striction type | Moving coil or ribbon type | Moving iron type | Stereo type

Low audio frequencies | High audio frequencies | Disk | Tape or film | Drum

Logic elements

Where two symbols are shown for a logic element, the second symbol is not recognised in B.S. 3939.

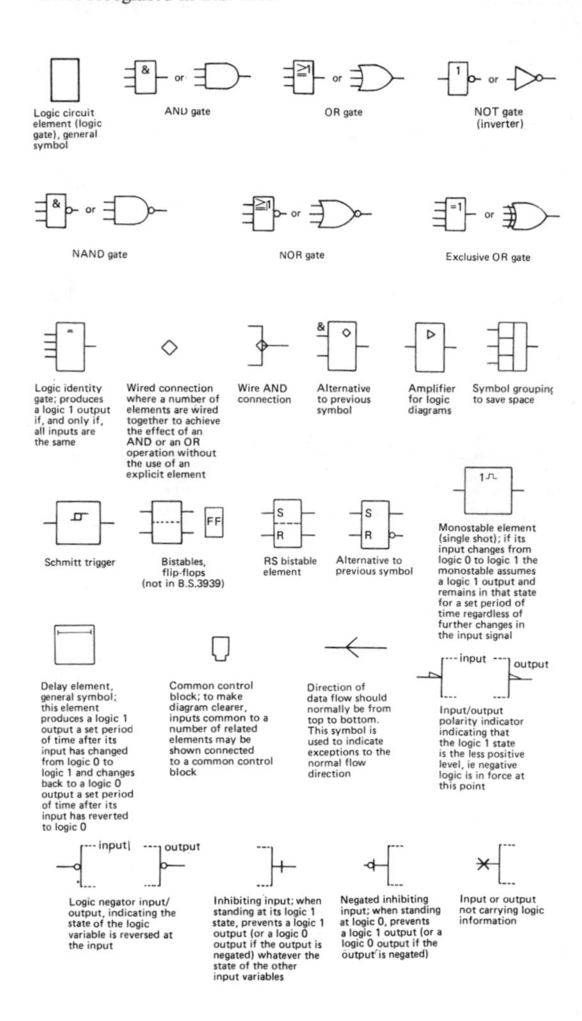

Block diagram symbols

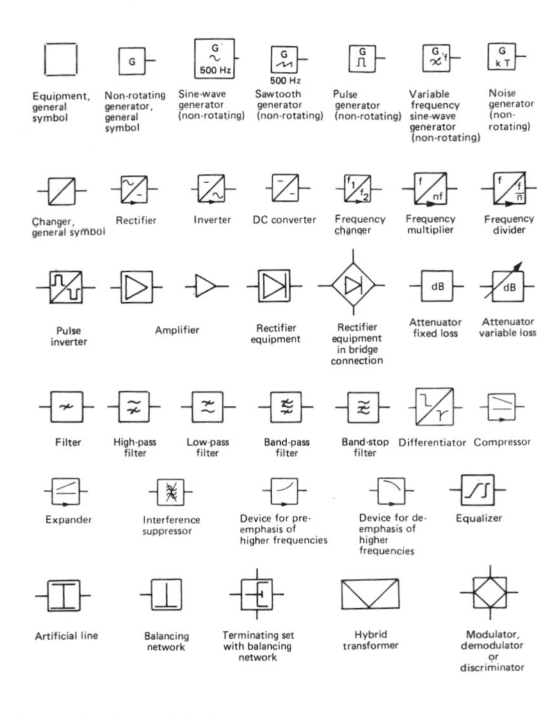

Frequency spectrum symbols

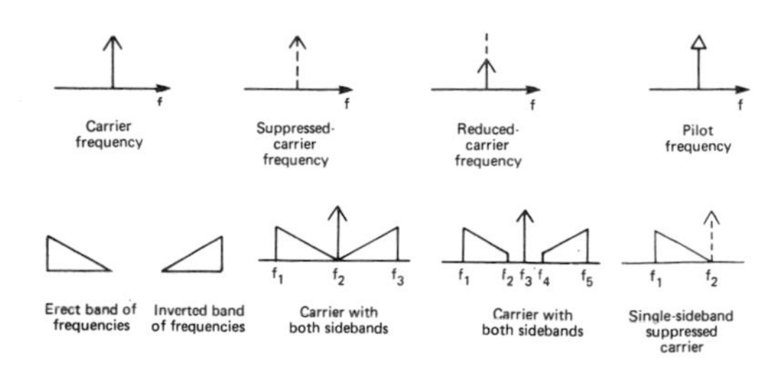

Op-amp data

Bipolar op-amps

Device type	*OP07*	*OP27*	*11*	*165*	*301A*	*308*	*324*	*348*	*531*
Supply voltage range (V_S)	±3 to ±18	±4 to ±18	±2·5 to ±20	±6 to ±18	±5 to ±18	±5 to ±18	3 to 32 ±1·5 to ±16	±10 to ±18	±5 to ±22
Max. differential input voltage	±30	±0·7	1	±15	30	30	32	24	15
Max. input V either input to earth	±22	±18	—	V_s	15	15	16	12	15
Operating temperature range	0–70	0–70	0–70	−40-150 (junc.)	0–70	0–70	0–70	0–70	0–70
Output short circuit duration	indef.	indef.	indef.	indef.	indef.	—	—	indef.	indef.
Max. total power dissipation		500	500	20(W)	500	500	625	500	300
Typical characteristics at 25°C, 2kΩ load									
Large signal open loop V. gain	132	123	109	80	88	102	100	96	96
Input resistance	33	4	10^5	500	2	40	10	2·5	20
Differential input offset voltage	0·06	0·03	0·2	2	2	10	2	1	2
Differential input offset current	0·8	12	0·001	20	3	1·5	5	4	50
Input bias current	±2·2	±15	0·04	200	70	<7	45	30	400
Common mode rejection ratio	120	100	130	70	90	100	70	90	100
Supply voltage rejection ratio	0·16	2	0·8	1000	16	16	—	15	10
Slew	0·17	2·8	0·3	6	0·4	—	—	0·6	35
Input offset voltage temp. coeff.	0·5	0·4	2	—	6	—	7	—	—
Input offset current temp. coeff.	12p	—	10f	—	20p	2p	10p	—	0·6n
Full power bandwidth	—	—	—	—	10	10	6	10	500
Output voltage swing	±13	±13	±12	24	±14	±13	28 or ±14	±12	±15

Op-amp data

F.E.T. op-amps

Max. ratings	*Device type*	*3130E*	*355*	*3140E 3240E*
	Package	*8-pin d.i.l.*	*8-pin d.i.l.*	*d.i.l.*
Supply voltage range (V_S)		+6 to +16V or ±3 to ±8V	±4 to ±18V	+4 to +36V or ±2 to ±18V
Max. differential input voltage		±8V	±30V	±8V
Max. input voltage either input with respect to earth		$\pm V_S$	$\pm V_S$	$\pm V_S$
Operating temperature range		0°–70°C	0°–70°C	0°–70°C
Max. total power dissipation		630mW	500mW	630mW
Output short circuit duration		indefinite	indefinite	indefinite
Typical characteristics at 25°C V_S = +15V				
Open loop voltage gain		110dB	106dB	100dB
Input resistance		$1{\cdot}5 \times 10^{12}\Omega$	$10^{12}\Omega$	$10^{12}\Omega$
Input offset voltage		8mV	3mV	5mV
Input offset current		0·5pA	10pA	0·5pA
Input bias current		5pA	30pA	10pA
Common mode rejection ratio		80dB	100dB	90dB
Supply voltage rejection ratio		300μV/V	10μV/V	100μV/V
Slew rate		10V/μs	5V/μs	9V/μs
Input offset voltage temp. coeff.		10μV/°C	5μV/°C	8μV/°C
Input bias current temp. coeff.				Doubles for every +20°C approx.
Full-power bandwidth		120kHz	60kHz	110kHz
Output voltage swing	R_L = 1kΩ	—	—	—
	R_L = 2kΩ	13V (V_S = 15V)	—	13V (V_S = 15V)
	R_L = 10kΩ	—	±13V	—

709	725CN	741 741N	741S	747	748	759	4558	5532	5534	5539	Units
±9 to ±18	±4 to ±22	±5 to ±18	±5 to ±18	±7 to ±18	±7 to ±18	7 to 36 ±3·5 to ±18	±3 to ±18	±3 to ±20	±3 to ±20	±8 to ±12	V
5	5	30	30	30	30	30	30	—	—	—	V
10	15	15	15	15	15	V_s	15	13	13	—	V
0–70	0–70	0–70	0–70	0–70	0–70	0–125 (junc.)	0–70	0–70	0–70	0–70	°C
5 sec.	5 sec.	indef.	indef.	indef.	indef.	indef.	indef.	indef.	indef.	—	
120	500	500	625	670	500	1300	680	1000	1000	550	mW (25°C)
93	127	106	100	106	106	106	109	100	100	52	dB
0·25	1·5	2	1	2	2	1·5	5	0·3	0·1	0·1	MΩ
2	2	1	2	1	2	1	0·5	0·5	0·5	2	mV
100	1·2	20	30	20	20	5	5	10	20	2000	nA
300	80	80	200	80	80	50	40	200	500	5000	nA
90	115	90	90	90	90	100	90	100	100	80	dB
25	20	30	10	30	30	10	30	10	10	200	µV/V
12	0·25	0·5	20	0·5	0·8	0·5	1	9	13	600	V/µs
3·3	2	5	3	—	—	—	—	—	—	—	µV/°C
0·1n	10p	0·5n	0·5n	0·5n	0·1n	—	—	—	—	—	A/°C
—	10	10	200	10	10	—	—	100	95	48000	kHz
±14	±10	±13	±13	±13	±13	±12·5	±13	±16	±16	+2·3 to −2·7	V

BIFET op-amps

351 353 8-pin d.i.l.	064 14-pin d.i.l.	071,072,074, 081,082,084 d.i.l.	091,092 d.i.l.
±5 to ±18V	±2 to ±18V	±3 to ±18V	3V to 36V
±30V	±30V	±30V	36V
±V_s	±V_s	±V_s	36V
0°–70°C	0°–70°C	0°–70°C	0°–70°C
500mW	680mW	680mW	1150mW
indefinite	indefinite	indefinite	indefinite
110dB	76dB	106dB	106dB
$10^{12}\Omega$	$10^{12}\Omega$	$10^{12}\Omega$	$10^{12}\Omega$
5mV	3mV	3mV (071,072,074) 5mV (081,082,084)	5mV
25pA	5pA	5pA	5nA
50pA	30pA	30pA	10nA
100dB	76dB	76dB	90dB
30µV/V	18µV/V	158µV/V	90dB
13V/µs	3·5V/µs	13V/µs	0·6V/µs
10µV/°C	20µV/°C	20µV/°C	10µV/°C
150kHz	40kHz	150kHz	9kHz
—	—	—	—
—	—	—	26V
±13·5V	±13·5V	±13·5V	27V

Logic terms

Astable Type of multivibrator circuit, producing a square wave oscillation.

Asynchronous Operation not dependent on clock pulses.

Bistable Type of multivibrator circuit, having two stable states.

Buffered Capable of driving external circuits, isolated from previous stage.

Clock Source of regular voltage pulses, used to synchronise systems.

Decoder Device capable of translating a BCD input to separate control line inputs (also known as a demultiplexer).

Dual Two, twin.

Edge triggered Operation of device takes place on rising (or falling) part of input pulse.

Enable Over-ride input.

Fan-out Number of devices that can be placed in parallel on output.

Flip-flop Two-state device, changes state when clocked.

Hex Six.

Latch Retains previous input state until over-ridden.

Monostable Type of multivibrator circuit, with one stable state.

Multiplexer Samples many inputs in sequence, gives one output.

Multivibrator Circuit having two output states, each of which may or may not be stable. Oscillators of astable, bistable, or monostable types can be built with multivibrators.

Octal Eight.

One-shot Gives single output pulse of defined duration from variable input pulse.

Open collector TTL output which needs external pull-up resistor, can be used to wire OR outputs.

Parity Check bit added to data, can be odd or even parity. In odd parity sum of data 1s + parity 1 is odd.

Propagation delay Time taken for signal to pass through a device, limits highest frequency of operation.

Quad Four.

Quiescent Stable state not driving a load.

Schmitt trigger Circuit with hysteresis.

Synchronous Operation dependent on clock pulses.

Basic logic symbols and truth tables

Logic symbols are to the Mil Std-806B specification, as they are in more general use than those of B.S.3939. Positive logic convention used, i.e. 1 = high, 0 = low.

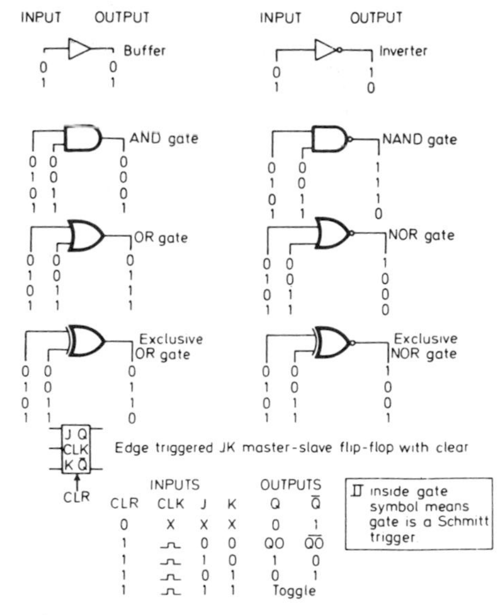

INPUTS				OUTPUTS	
CLR	CLK	J	K	Q	Q̄
0	X	X	X	0	1
1	⎍	0	0	QO	$\overline{QO}$
1	⎍	1	0	1	0
1	⎍	0	1	0	1
1	⎍	1	1	Toggle	

⎍ High level pulse, data is transferred on falling edge of pulse.

QO The level of Q before indicated output conditions were established.

Toggle Each output changes to its complement on each active transition (pulse) of clock.

Medium scale integrated logic symbols and terminology

Medium scale integrated (MSI) logic elements are represented by rectangular blocks with appropriate external AND/OR gates when necessary. A small circle at an external input means that the specific input is active Low, i.e., it produces the desired function, in conjunction with other inputs, if its voltage is the lower of the two logic levels in the system. A circle at the output indicates that when the function designated is True, the output is Low. Generally, inputs are at the top and left and outputs appear at the bottom and right of the logic symbol. An exception is the asynchronous Master

Reset in some sequential circuits which is always at the left-hand bottom corner.

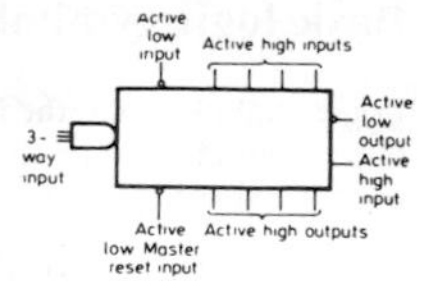

Inputs and outputs are generally labelled with mnemonic letters. Those used in this book are listed below. Note that an active Low function labelled outside of the logic symbol is given a bar over the label, while the same function inside the symbol is labelled without the bar. When several inputs or outputs use the same letter, subscript numbers starting with zero are used in an order natural for device operation.

Label	*Meaning*
A, B, C, D etc.	Data inputs (binary weighted where applicable: A=1; B=2; C=4; D=5 etc.)
a, b, c, d etc.	Segment outputs of a seven-segment decoder driver
BCD	Binary coded decimal
BI	Blanking input
$C_{in, out}$	Carry in or out. Sometimes may be labelled CI, CY
CE	Clock enable
CF	Cascade feedback
CEP	Count enable parallel input
CER	Count enable ripple input
CK	Clock input
CP	Clock pulse, generally a high-to-low transition. An active high clock (no circle) means outputs change on low-to-high clock transition
CS	Chip select
D, J, K, R, S	Data inputs to JK, SR and D flip-flops, latches, registers and counters
DIS	Disable 3-state output
EN	Enable device, generally active low
GND	Ground (0 volts) terminal (TTL)
I/O	Input/output
INC	Increment
INH	Inhibit
LE	Latch enable
LT	Lamp test
MR	Master reset, asynchronously resets all outputs to zero, overriding all other inputs. Generally active low
OEN	Output enable
OF	Overflow
PE	Parallel enable, a control input used to synchronously load information in parallel into an otherwise autonomous circuit (generally active low)
PH	Phase input for liquid crystal displays
P/S	Parallel/serial mode control input

$Q, \bar{Q}$ — General term, and complement, for output of sequential circuit. May have a letter indicating weighting
QP — Phase pulse output
R — Reset
RBI — Ripple blanking input
RBO — Ripple blanking output
RC, C, R — Capacitor and resistor timing on monostables
RCO — Ripple carry output
S — Preset input of flip-flop
$S_{1,\ 2\ etc}$ — Sum outputs
$S_{in,\ out}$ — Serial inputs, outputs of shift register
SDL — Serial data in left shift
SDR — Serial data in right shift
SF — Source follower output
SQ — Serial output
SR — Synchronous reset
ST — Strobe input
T — Trigger input
TC — Terminal count output (1111 for up binary counters, 1001 for up decimal counters, or 0000 for down counters)
U/D — Up/down mode control input
VCO — Voltage controlled oscillator
VI — Input to voltage controlled oscillator
VO — Voltage controlled oscillator output
V_{CC} — Positive supply terminal
V_{DD} — Positive supply terminal (CMOS)
V_{EE} — Negative supply terminal (CMOS)
V_{SS} — 0 volts supply terminal (CMOS)
W — User-selected positive or negative logic
WE — Write enable
X — Data inputs to selector
Z, O, F — General term for outputs of combinational circuits. May have a letter indicating weighting
⎍ — Schmitt trigger device or function

Comparison of logic families

Since RTL was introduced in the early 1960s, there has been a steady progression in technology; the design engineer now has a wide choice of integrated circuit ranges and operating parameters.

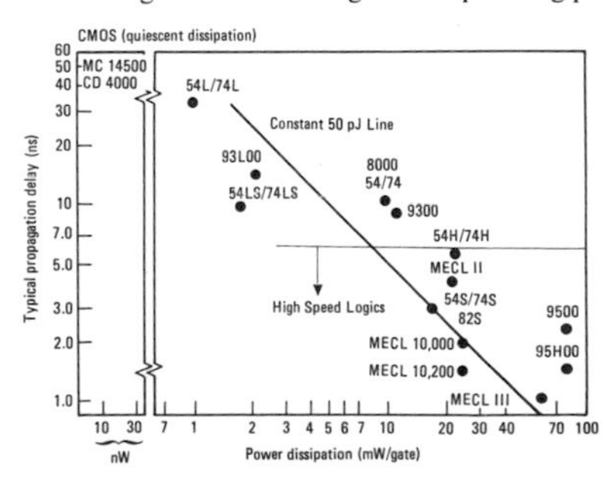

Speed/power characteristics of major logic lines

It is apparent that speed/power comparisons are not sufficient in themselves. Other important parameters to be considered are noise immunity, supply voltage requirements and fan out.

Logic family	Noise immunity Volts	Prop delay[1] ns	Fan out	Max. toggle speed[1] MHz
DTL[2]	0·3	30	8	4
RTL[2]	0·3	12	5	1·5
74 series	0·04	9	10	15
74H series	0·4	6	10	40
74S series	0·3	3	10	125
74LS series	0·3	9	10	25
74C series		30	>50	10
4000 series	4·5	30	>50	10

[1]Typical.
[2]Not recommended for new designs.

Power supply requirements
Each logic type has different power supply requirements and since system economics can be greatly affected by the cost of power supplies it is important to establish exact power supply parameters.

Logic family	Supply voltage Nominal V	Min. V	Max. V	Power diss. per package mW (typ)	Decoupling and other requirements
RTL	3·6	3·24	3·96	20	No special precautions
DTL	5·0	4·5	5·5	30	No special precautions
74 series	5·0	4·75	5·25	40	0·1 μF decoupling capacitor for every 8 packages
74H series	5·0	4·75	5·25	60	
74S series	5·0	4·75	5·25	40	
74LS series	5·0	4·75	5·25	8	
74C series	5·0	4·75	5·25	2	
4000 series	—	3·0	18·0	0·01	No special precautions

With the exception of the 74C series of devices, all series in the 74 family are of transistor-transistor-logic (TTL) construction. The 74C series (which is of CMOS construction) is pin compatible with all 74 family members and therefore the generic TTL title usually includes it.

TTL 74 family and CMOS 4000 family devices have generally superseded all other logic types.

TTL data

All devices are listed by generic family number, e.g. 7400. Certain devices may only be available, however, as members of as few as one series of the 74 family. Readers are referred to relevant data books for further details.

Selection by device number

Device	Description
7400	Quad 2-input Positive NAND Gate
7401	Quad 2-input Positive NAND Gate (open collector o/p)
7401A	Quad 2-input Positive NAND Gate (open collector o/p)
7402	Quad 2-input Positive NOR Gate (open collector o/p)
7403	Quad 2-input Positive NAND Gate (open collector o/p)
7404	Hex Inverter
7405A	Hex Inverter (open collector o/p)
7406	Hex Inverter/Buffer 30V o/p
7407	Hex Buffer 30V o/p
7408	Quad 2-input Positive AND Gate
7409	Quad 2-input Positive AND Gate
7410	Triple 3-input Positive NAND Gate
7411	Triple 3-input AND Gate
7412	Triple 3-input NAND Gate (open collector o/p)
7413	Dual 4-input Schmitt Trigger
7414	Schmitt Hex Inverter Buffer
7415	Triple 3-input AND Gate with Open Collector Output
7416	Hex Inverter/Buffer 15V o/p
7417	Hex Buffer 15V o/p
7420	Dual 4-input Positive NAND Gate
7421	Dual 4-input AND Gate
7422	Dual 4-input NAND Gate with Open Collector Output
7425	Dual 4-input NOR Gate with Strobe
7426	Quad 2-input High Voltage Interface NAND Gate
7427	Triple 3-input NOR Gate
7428	Quad 2-input NOR Buffer (Fan Out 30)
7430	8-input Positive NAND Gate
7432	Quad 2-input OR Gate
7433A	Quad 2-input NOR Buffer 15V
7437	Quad 2-input NAND Buffer
7438A	Quad 2-input NAND Buffer 15V
7440	Dual 4-input Buffer NAND Gate
7441A	BCD-to-Decimal Decoder/Nixie Driver
7442	BCD-to-Decimal Decoder
7445	BCD-to-Decimal Decoder/Driver 30V output o/c
7446A	BCD-to-Seven Segment Decoder/Driver 30V/40mA
7447	BCD-to-Seven Segment Decoder/Driver 15V/20mA
7447A	BCD-to-Seven Segment Decoder/Driver 15V/40mA
7448	BCD-to-Seven Segment Decoder/Driver
7449	BCD-to-7-segment driver with Open Collector Output
7450	Expandable Dual 2 wide, 2 i/p AND-OR-INVERT Gate
7451	Dual 2 wide, 2 i/p AND-OR-INVERT Gate
7453	Expandable 4 wide, 2 i/p AND-OR-INVERT Gate
7454	4 wide, 2 i/p AND-OR-INVERT Gate
7455	2 wide, 4 i/p AND-OR-INVERT Gate
7460	Dual 4-input Expander
7464	4-2-3-2-input AND-OR-Invert Gate
7470	Positive Edge-triggered J-K Flip Flops
7472	J-K Master-Slave Flip Flops (AND inputs)
7473	Dual J-K Master-Slave Flip Flops
7474	Dual D-Type Edge Triggered Flip Flops
7475	4-bit bistable latch = Quad bistable latch
7476	Dual J-K Master-Slave Flip Flops + preset and clear
7478	Dual J-K Flip-flop with Preset, Common Clear and Clock

7481	16-bit Active Element Memory
7482	2-bit Binary Full Adder
7483A	4-bit Full Adder with Carry
7484	16-bit Active Element Memory
7485	4-bit Comparator
7486	Quad 2-input Exclusive OR Gate
7489	64-bit RAM (16 × 4W)
7490	Decade Çounter
7491	8-bit Shift Registers
7492	Divide-by-twelve Counter
7493	4-bit Binary Counter
7494	4-bit Shift Registers (Parallel-In, Serial-Out)
7495	4-bit Right Shift, Left Shift Register
7496	5-bit Shift Registers (Dual Para-In, Para-Out)
74100	8-bit Bistable Latch
74107	Dual J-K Master Slave Flip Flop
74109	Dual Positive Edge Triggered Flip-flop with Preset and Clear
74112	Dual Negative Edge Triggered J-K Flip-flop with Preset and Clear
74113	Dual Negative Edge Triggered J-K Flip-flop with Preset
74114	Dual Negative Edge Triggered J-K Flip-flop with Preset and Clear
74121	Monostable Multivibrator
74122	Monostable Multivibrator with reset
74123	Dual Monostable Multivibrator with reset
74124	Universal Pulse Generator
74125	Quad Buffer with 3-state Active Low Enable Output
74126	Quad Buffer with 3-state Active High Enable Output
74128	Quad Line Driver
74132	Quad 2-input Schmitt NAND
74133	13-input NAND
74137	Demultiplexer
74138	3 line to 8 line Decoder/Demultiplexer
74139	Dual 2-to-4 Line Multiplexer
74141	BCD-to-Decimal Decoder/Driver
74145	BCD-to-Seven Segment Decoder/Driver 15V output
74147	10-line Priority Decimal to 4-line BCD Priority Encoder
74148	8-to-3 Octal Priority Encoder
74150	16-bit Data Selector
74151	8-bit Data Selector (with strobe)
74153	Dual 4 to 1 line Data Selector 1 MPX
74154	4 line to 16 line Decoder
74155	Dual 2-to-4 line Decoder/DeMPX (totem pole output)
74156	Dual 2-to-4 line Decoder/DeMPX (open collector output)
74157	Quad 2 line to 1 line Selector
74158	Quad 2-input Inverting Multiplexer
74160	Synchronous Decade Counter
74161	Asynchronous Binary Counter with Reset
74162	Synchronous Decade Counter
74163	Synchronous Binary Counter
74164	8-bit Shift Register, Serial In-Parallel Out
74165	8-bit Shift Register, Parallel In-Serial Out
74169	4-stage Synchronous Bidirectional Counter
74173	4-bit D-Type Register
74174	Hex Type 'D' Flip Flop
74175	Quad 'D' Flip Flop with common reset
74180	8-bit Odd/Even Parity Generator/Checkers

74181	4-bit Arithmetic Logic Out
74182	Carry-Look-Ahead Unit
74190	Synchronous Up/Down Decade Counter (Single Clock Unit)
74191	Synchronous Up/Down 4-bit Binary Counter (Single Clock Unit)
74192	Synchronous 4-bit Up/Down Counter
74193	Synchronous 4-bit Up/Down Counter
74194	4-bit Universal Shift Register
74195	Synchronous 4-bit Parallel Shift Register with J-K inputs
74196	50Mhz Presettable Decade Counter/Latch (Bi-Quinary)
74197	4-bit Presettable Ripple Counter
74200	256-bit Random Access Memory (RAM)
74221	Dual Monostable Multivibrator
74240	Octal Inverting Buffer with 3-state Outputs
74241	Octal Buffer with 3-state Outputs
74242	Octal Bus Inverting Transceiver
74243	Octal Bus Transceiver
74244	Octal Buffer with 3-state Outputs
74245	Octal Bus Transceiver with 3-state Outputs
74251	Selector Multiplexer with 3-state Outputs
74253	Dual 4-input Multiplexer with 3-state Output
74256	Dual 4-bit Addressable Latch
74257	Quad 2-input Multiplexer with 3-state Output
74258	Quad 2-input Multiplexer with Inverting 3-state Output
74259	8-bit Addressable Latch
74273	8-bit Register with Clear
74280	9-bit Parity Generator/Checker
74283	4-bit Full Adder with Carry
74298	Quad 2-port Register
74299	8-bit Universal Storage Shift Register with 3-state Output
74321	Crystal Oscillator
74323	8-bit Universal Storage Shift Register with 3-state Output
74352	Dual 4-bit Inverting Multiplexer
74353	Dual 4-bit Multiplexer with 3-state Inverting Output
74354	Transparent Data Selector Multiplexer
74356	Data Selector Multiplexer
74365	Hex Buffer with 2-input NOR Enable
74366	Hex Inverting Buffer with 2-input NOR Enable
74367	Hex Buffer with 3-state Output
74368	Hex Inverting Buffer with 3-state Output
74373	Octal Latch with 3-state Output
74374	Octal D-type Flip-flop with 3-state Output
74378	Hex D-type Flip-flop
74381	4-bit Arithmetic Logic Unit
74390	Dual Decade Counter
74393	Dual 4-bit Binary Counter
74395	4-bit Cascadable Shift Register
74399	Quad, 2-part Register
74423	Retriggerable Monostable Multivibrator
74442	Quad Tridirectional Transceiver
74443	Quad Tridirectional Inverting Transceiver
74444	Quad Tridirectional Transceiver
74533	Inverting Octal D-Type Latch
74534	Inverting Octal D-type Flip Flop
74563	Octal Transparent Latch with Inverted Outputs
74564	Octal Edge-Triggered Flip Flop with Inverted Outputs
74620	Octal Bus Transceiver

74625 Voltage Controlled Oscillator
74655 Inverting Octal Buffer/Line Driver with 3-state Outputs
74657 Octal Bi-directional Transceiver with Parity
74669 4-bit Binary Counter
74670 4 × 4 Register File with 3-state Output
74673 16-bit Serial to Parallel Shift Register
74674 16-bit Parallel to Serial Shift Register
74682 8-bit Magnitude Comparator
74688 8-bit Magnitude Comparator with Totem Pole Output
741242 Quad Bus Transceiver – Inverting
741243 Quad Bus Transceiver – Non-inverting
744002 Dual 4-input NOR gates
744017 Decade Counter Divider
744020 14-bit Binary Counter
744040 12-bit Binary Counter
744049 Hex-Inverter Buffer
744050 Hex Buffer
744060 14-bit Binary Counter
744075 Triple 3-Input OR Gate
744078 8-input NOR Gate
744511 BCD-Seven Segment Latch/Decoder/Driver
744514 4-bit Latch to 1-of-16 Decoder
744538 Dual Precision Retriggerable/Resettable Monostable Multivibrator
744543 BCD- to-Seven Segment Latch/Decoder/Driver

Selection by function

Gates

AND
- Quad 2-input **7408**
- Quad 2-input open collector o/p **7409**
- Triple 3-input **7411**
- Triple 3-input open collector o/p **7415**
- Dual 4-input **7421**

OR
- Triple 3-input **744075**
- Quad 2-input **7432**

Exclusive OR
- Quad 2-input **7486**

NAND
- Quad 2-input **7400**
- Quad 2-input open collector o/p **7401**
- Quad 2-input open collector o/p **7403**
- Triple 3-input **7410**
- Dual 4-input **7420**
- Dual 4-input open collector o/p **7422**
- Quad 2-input high voltage **7426**
- 8-input **7430**
- Quad 2-input buffer **7437**
- Dual 2-input open collector o/p **7438**
- Dual 4-input buffer **7440**
- 13-input **74133**

NOR
- Quad 2-input **7402**
- Dual 4-input **744002**
- Dual 4-input with strobe **7425**
- Triple 3-input **7427**

Quad 2-input buffer	**7428**
Quad 2-input buffer	**7433**
Quad 2-input exclusive	**74266**
8-input	**744078**
Schmitt	
Dual 4-input NAND	**7413**
Hex inverting	**7414**
Quad 2-input NAND	**74132**
AND-OR-Invert	
Dual 2-wide, 2-input	**7451**
4-wide	**7454**
2-wide, 4-input	**7455**
4-2-3-2-Input	**7464**
Buffers	
Hex	**744050**
Hex inverting	**744049**
Hex inverting	**7404**
Hex inverting open collector o/p	**7405**
Hex inverting open collector o/p	**7406**
Hex open collector o/p	**7407**
Hex inverting open collector o/p	**7416**
Quad 3-state active low enable	**74125**
Quad 3-state active high enable	**74126**
Hex 2-input NOR enable	**74365**
Hex inverting, 2-input NOR enable	**74366**
Hex 3-state	**74367**
Hex 3-state inverting	**74368**

Line/bus, drivers/transceivers

Quad line driver	**74128**
Octal buffer 3-state inverting	**74240**
Octal buffer 3-state	**74241**
Quad bus transceiver inverting	**74242**
Quad bus transceiver	**74243**
Quad bus transceiver, inverting	**741242**
Quad bus transceiver	**741243**
Octal buffer 3-state	**74244**
Octal bus transceiver 3-state	**74245**
Quad tridirectional transceiver true	**74442**
Quad tridirectional transceiver inverting	**74443**
Quad tridirectional transceiver	**74444**
Octal bus transceiver	**74620**
Octal bus transceiver	**74640**
Octal bus transceiver	**74643**
Octal, bi-directional transceiver with parity	**74657**
Octal buffer/line driver, inverting, 3-state o/p	**74655**
Octal, buffer/line drive, non-inverting, 3-state o/p	**74656**

Flip flops (bistables)

D-type	
Dual edge triggered	**7474**
4-bit	**7475**
Hex with clear	**74174**
Inverting octal	**74534**
Quad with clear	**74175**
Octal 3-state	**74374**
Octal common enable	**74377**

Octal, edge-triggered, inverted outputs	**74564**
Hex	**74378**
Octal transparent latch	**74573**
Octal transparent latch inverted	**74580**
J-K	
AND gated positive edge triggered	**7470**
AND gated master slave	**7472**
Dual with clear	**7473**
Dual with preset and clear	**7476**
Dual with preset, common clear and clock	**7478**
Dual with clear	**74107**
Dual positive edge triggered preset and clear	**74109**
Dual negative edge triggered preset and clear	**74112**
Dual negative edge triggered preset	**74113**
Dual negative edge triggered preset and clear	**74114**
Monostable multivibrators	
Single	**74121**
Dual retriggerable with clear	**74123**
Dual retriggerable/resettable	**744538**
Dual retriggerable	**74423**
Dual	**74221**
Latches	
Dual 4-bit addressable	**74256**
Inverting octal D-type	**74533**
4-bit to 1-of-16 decoder	**744514**
8-bit addressable	**74259**
8-bit register with clear	**74273**
Quad 2-port register	**74298**
Octal 3-state	**74373**
Octal transparent, inverted outputs	**74563**

Arithmetic functions

4-bit full adder with carry	**7483A**
4-bit magnitude comparator	**7485**
4-bit arithmetic logic unit	**74181**
4-bit arithmetic logic unit	**74381**
4-bit full adder with carry	**74283**
4 × 4 register file 3-state	**74670**
8-bit magnitude comparator	**74682**
8-bit magnitude comparator, totem-pole outputs	**74688**

Counters

Decade up	**7490**
Divide by 12	**7492**
4-bit binary	**7493**
B.C.D asynchronous reset	**74160**
Binary asynchronous reset	**74161**
B.C.D. synchronous reset	**74162**
Binary synchronous reset	**74163**
Binary up/down synchronous	**74191**
Decade up/down synchronous	**74192**
Binary up/down synchronous with clear	**74193**
Decade presettable ripple	**74196**
4-bit presettable ripple	**74197**
Dual decade	**74390**
Dual 4-bit binary	**74393**
4-bit binary	**74669**
4-stage synchronous up/down	**74169**

Decade counter/divider	**744017**
4-bit binary	**744020**
12-bit binary	**744040**
14-bit binary	**744060**
Shift registers	
4-bit	**7495**
5-bit	**7496**
8-bit serial in parallel out	**74164**
8-bit parallel to serial	**74165**
4-bit universal	**74194**
4-bit parallel access	**74195**
4-bit D-type	**74173**
8-bit universal storage 3-state	**74299**
8-bit universal storage 3-state	**74323**
4-bit cascadable	**74395**
16-bit serial to parallel	**74673**
16-bit parallel to serial	**74674**
Quad, 2-part	**74399**
Encoders, decoders/drivers	
Decoders	
B.C.D.–decimal	**7442**
B.C.D.–decimal driver	**7445**
B.C.D.–7-segment driver open collector o/p	**7447**
B.C.D.–7-segment driver	**7448**
B.C.D.–7-segment driver open collector o/p	**7449**
B.C.D.-to-7-segment latch/decoder/driver	**744511**
B.C.D.-to-7-segment latch/decoder/driver	**744543**
De-multiplexer	**74137**
3-to-8 line multiplexer	**74138**
Dual 2-to-4 line multiplexer	**74139**
B.C.D.–decimal driver	**74141**
B.C.D.–decimal driver	**74145**
4-to-16 line	**74154**
Dual 1-of-4	**74155**
Dual 1-of-4 open collector o/p	**74156**
Encoders/multiplexers	
Octal priority encoder 8-to-3	**74148**
8-input multiplexer	**74151**
Dual 4-input multiplexer	**74153**
Quad 2-input multiplexer	**74157**
Quad 2-input multiplexer inverting	**74158**
Parity generator/checker 9-bit odd/even	**74180**
Selector multiplexer 3-state	**74251**
Dual 4-input multiplexer 3-state	**74253**
Quad 2-input multiplexer 3-state	**74257**
Quad 2-input multiplexer 3-state inverting	**74258**
Dual 4-input multiplexer inverting	**74352**
Dual 4-input multiplexer 3-state inverting	**74353**
Data selector multiplexer transparent	**74354**
Data selector multiplexer	**74356**
10-line decimal to 4-line B.C.D.	**74147**
Miscellaneous	
Crystal oscillator	**74321**
Voltage controlled oscillator	**74625**
9-bit parity generator/checker	**74280**

TTL pinouts

7400

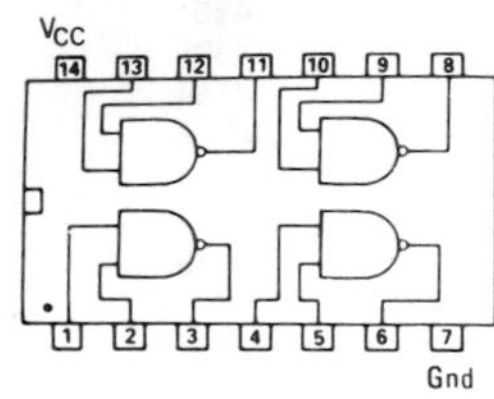

7401

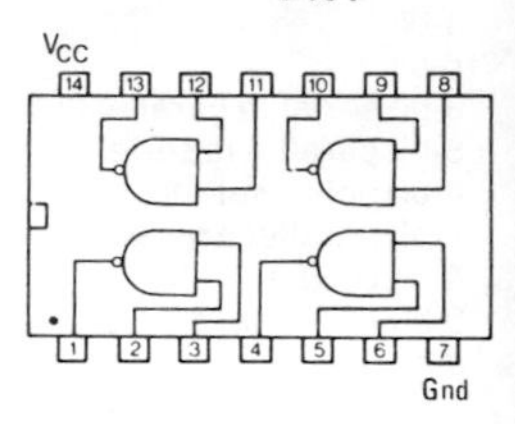

7402

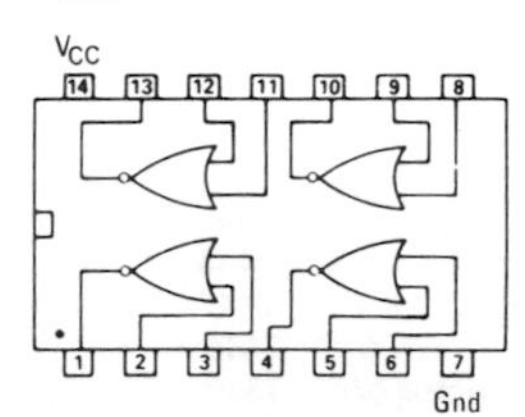

7403

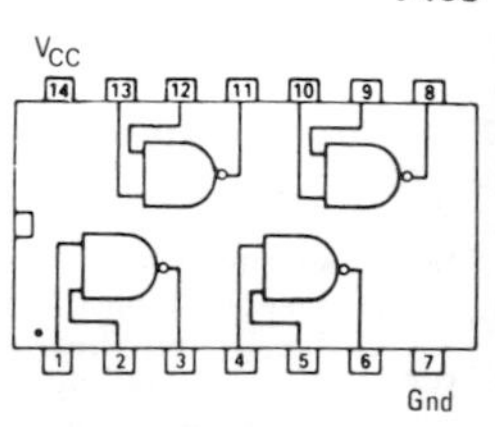

7404

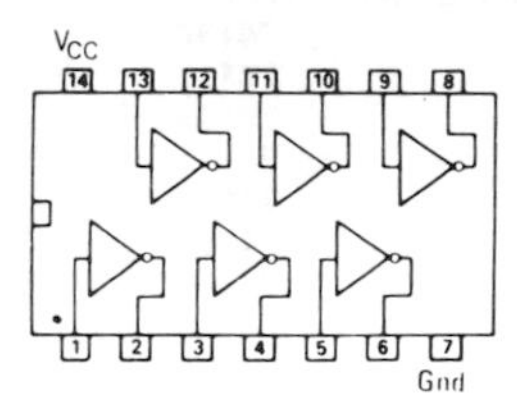

7405

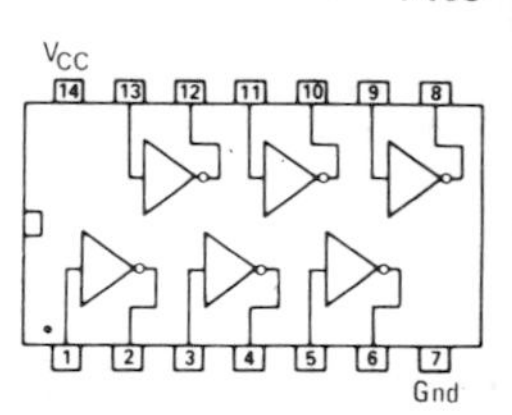

7406

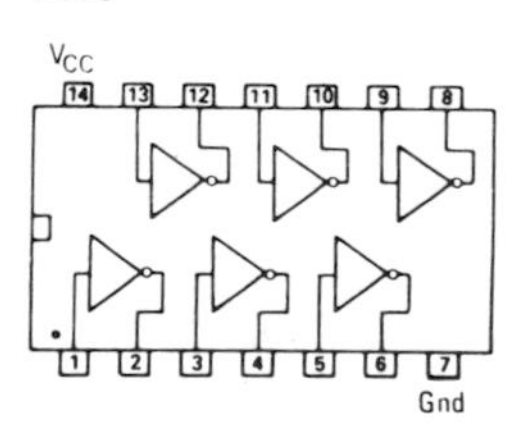

7407

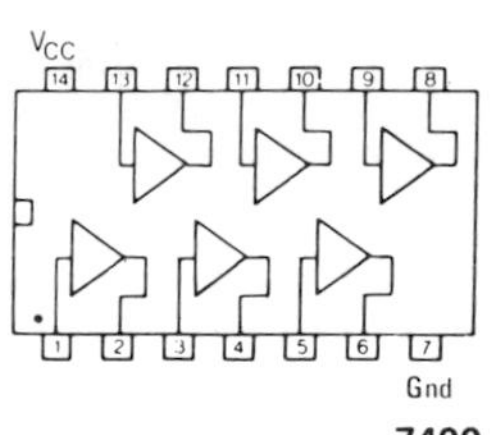

7408

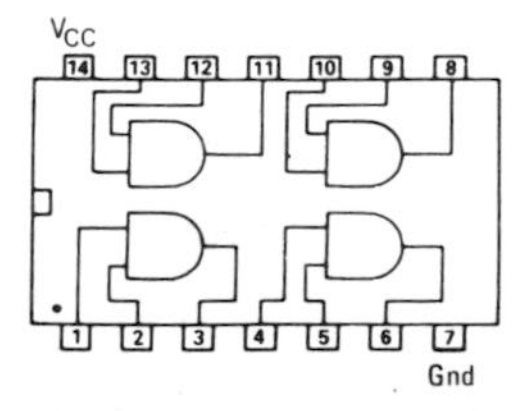

7409

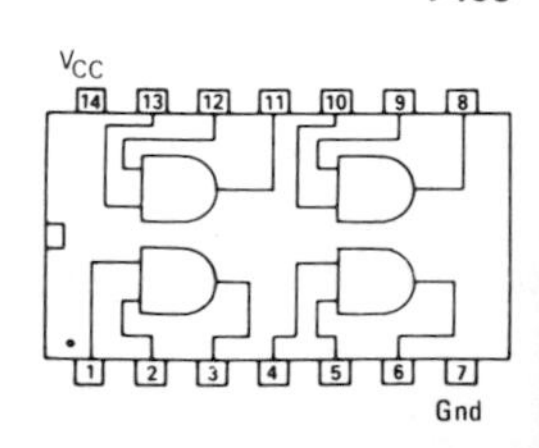

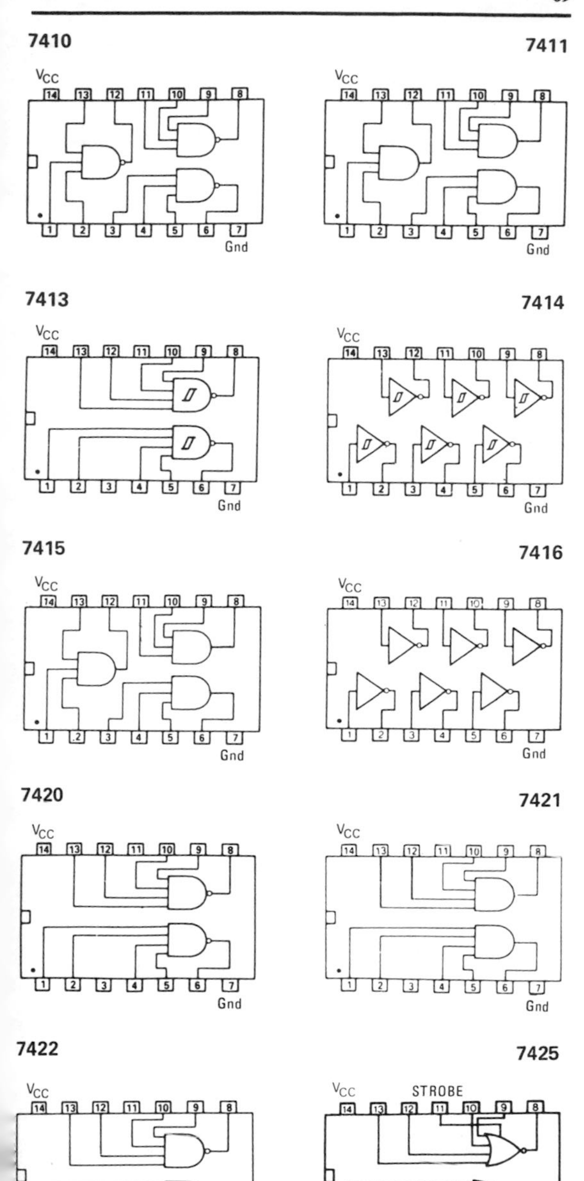
7410
7411
7413
7414
7415
7416
7420
7421
7422
7425
VCC
Gnd
STROBE
STROBE

7426

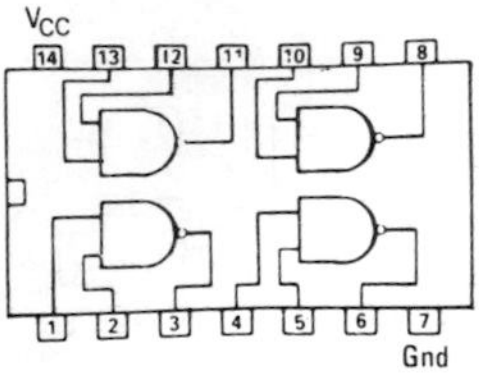

Vcc 14 13 12 11 10 9 8
1 2 3 4 5 6 7 Gnd

7427

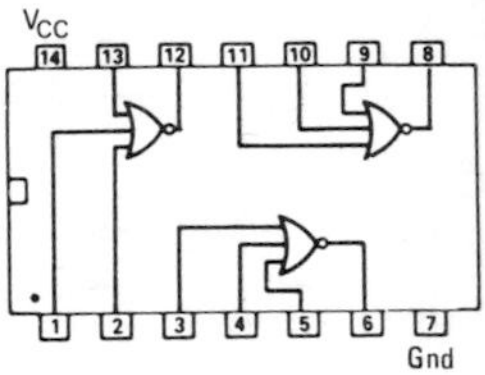

Vcc 14 13 12 11 10 9 8
1 2 3 4 5 6 7 Gnd

7428

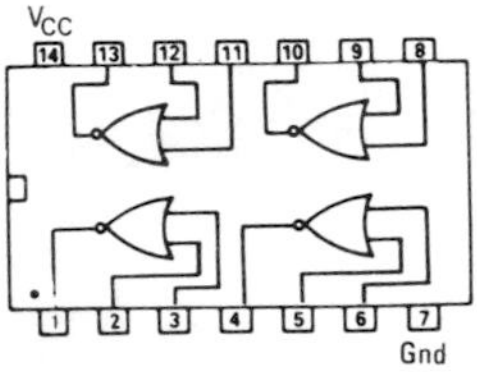

Vcc 14 13 12 11 10 9 8
1 2 3 4 5 6 7 Gnd

7430

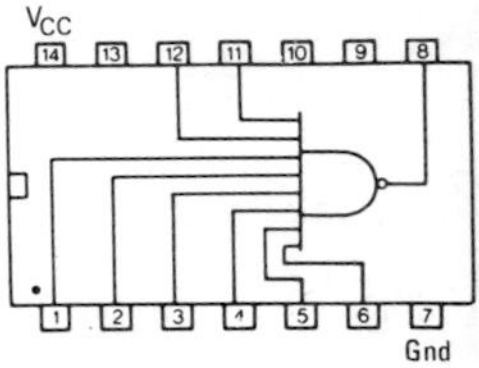

Vcc 14 13 12 11 10 9 8
1 2 3 4 5 6 7 Gnd

7432

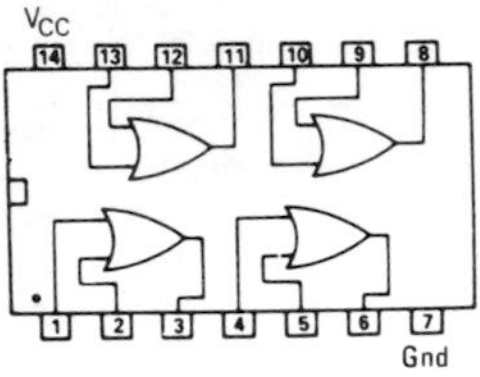

Vcc 14 13 12 11 10 9 8
1 2 3 4 5 6 7 Gnd

7433

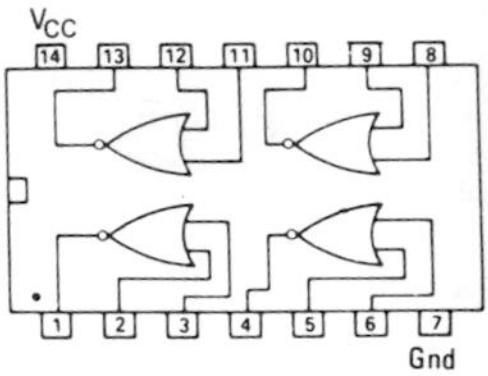

Vcc 14 13 12 11 10 9 8
1 2 3 4 5 6 7 Gnd

7437

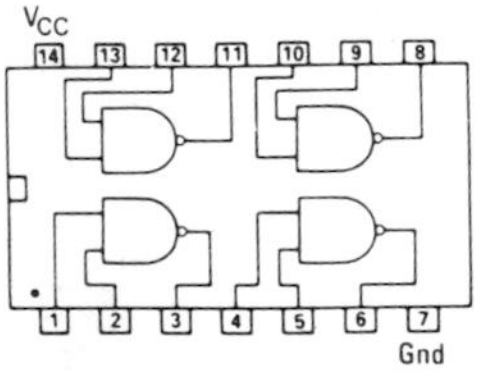

Vcc 14 13 12 11 10 9 8
1 2 3 4 5 6 7 Gnd

7438

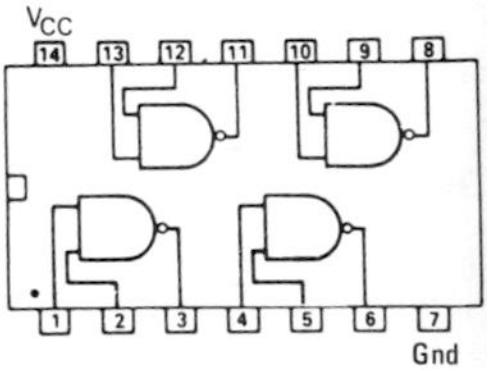

Vcc 14 13 12 11 10 9 8
1 2 3 4 5 6 7 Gnd

7440

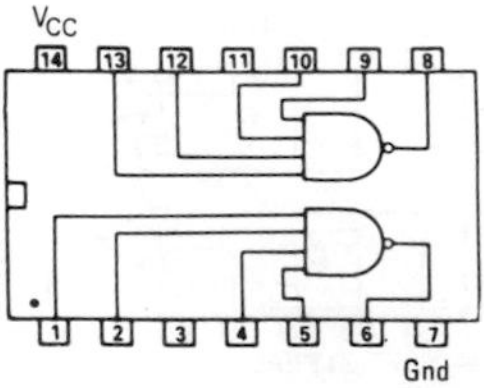

Vcc 14 13 12 11 10 9 8
1 2 3 4 5 6 7 Gnd

7442

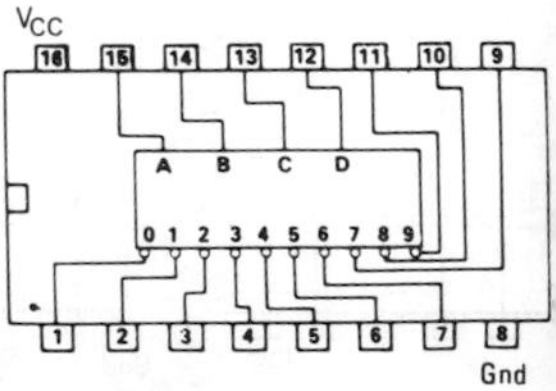

Vcc 16 15 14 13 12 11 10 9
A B C D
0 1 2 3 4 5 6 7 8 9
1 2 3 4 5 6 7 8 Gnd

7445

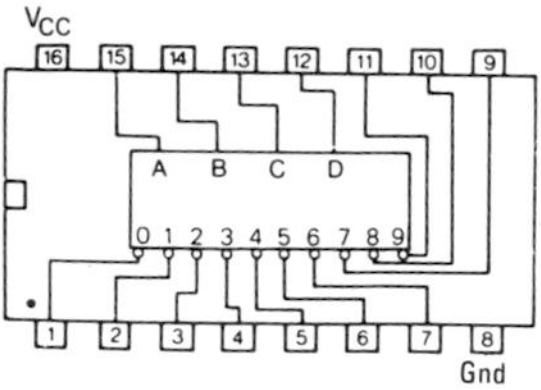

7447

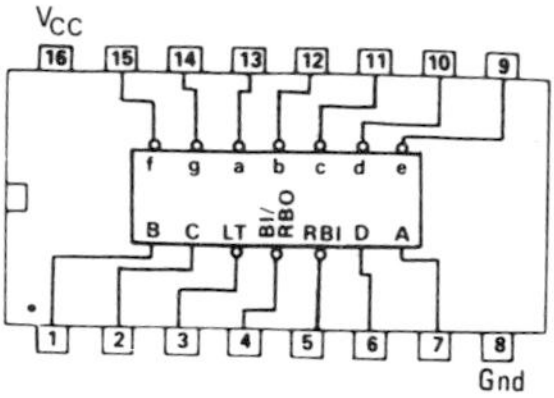

7448

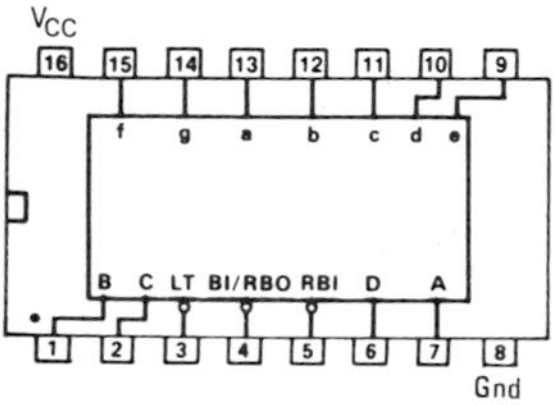

7449

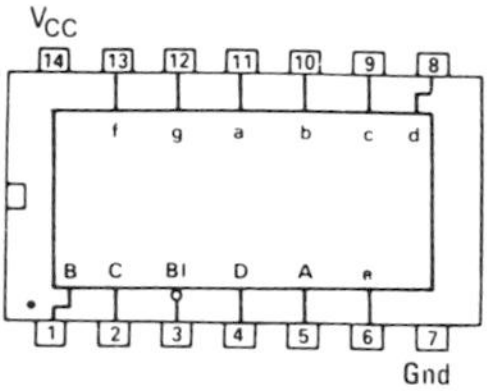

7451

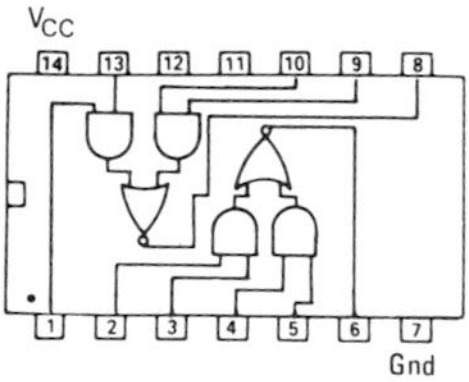

7454

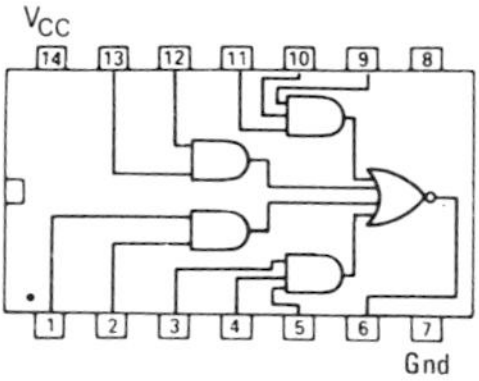

7455

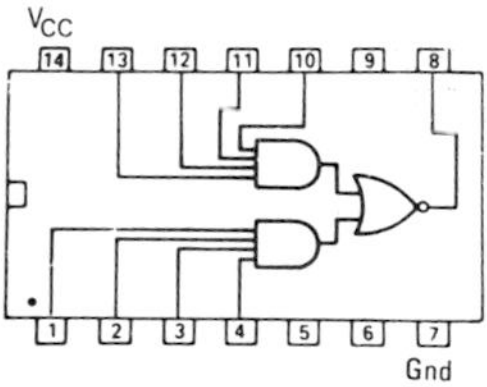

7464

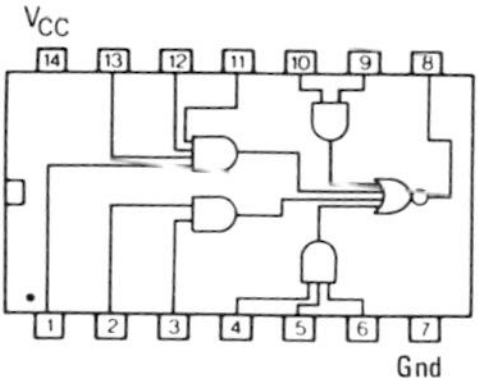

7470

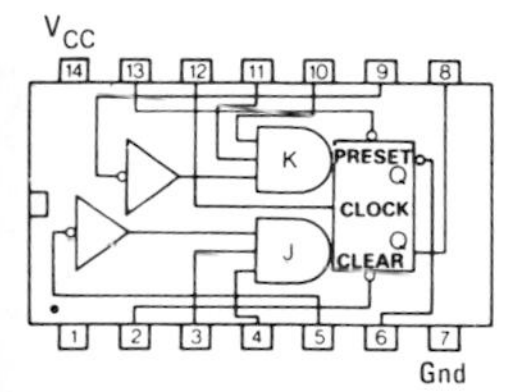

7472

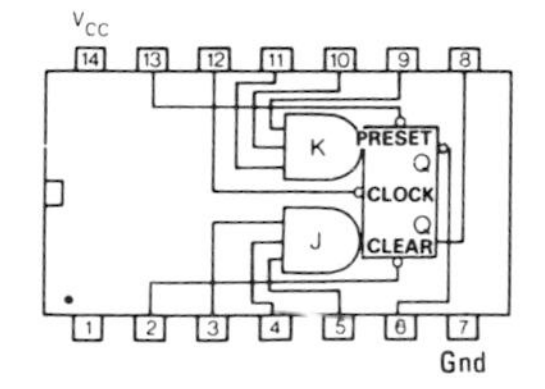

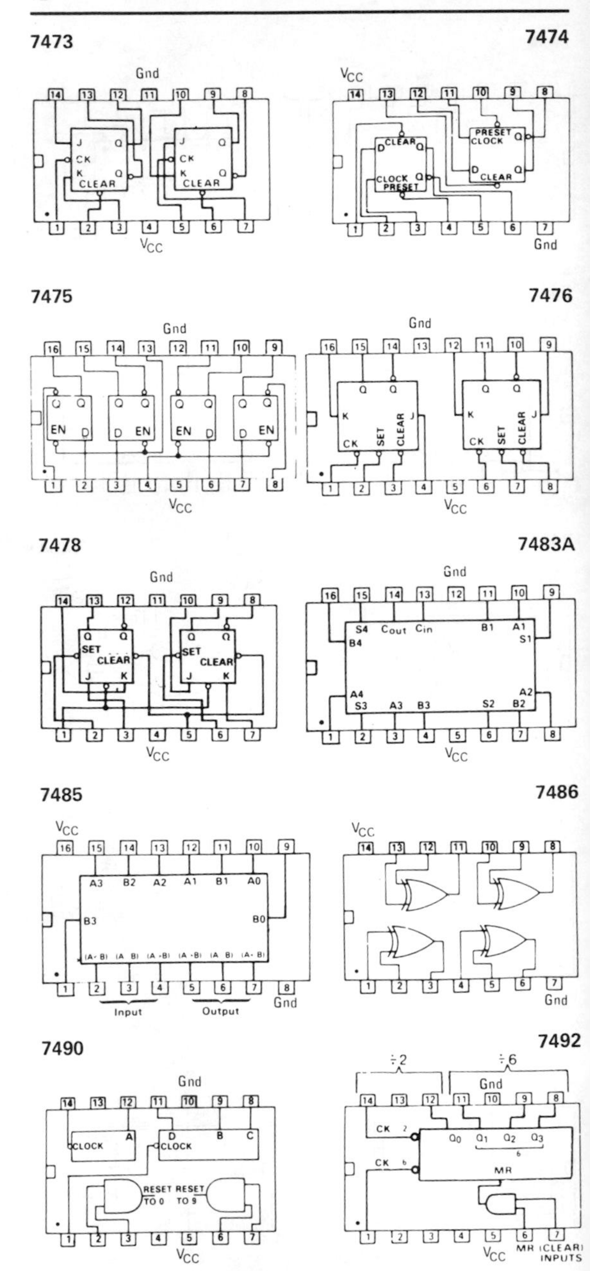
7473
Gnd
J Q
CK
K Q
CLEAR
V_{CC}
7474
V_{CC}
PRESET
CLOCK Q
D Q
CLEAR
CLOCK Q
PRESET
Gnd
7475
Gnd
Q Q
EN D
D EN
V_{CC}
7476
Gnd
Q Q
K J
CK SET CLEAR
V_{CC}
7478
Gnd
Q Q
SET CLEAR
J K
V_{CC}
7483A
Gnd
S4 C_{out} C_{in} B1 A1
B4 S1
A4 A2
S3 A3 B3 S2 B2
V_{CC}
7485
V_{CC}
A3 B2 A2 A1 B1 A0
B3 B0
Input
Output
Gnd
7486
V_{CC}
Gnd
7490
Gnd
CLOCK A
D CLOCK B C
RESET TO 0
RESET TO 9
V_{CC}
7492
÷2
÷6
Gnd
CK 2
CK 6
Q_0 Q_1 Q_2 Q_3
MR
V_{CC}
MR (CLEAR) INPUTS

7493

7495B

7496

74107

74109

74112

74113

74114

74121

74123

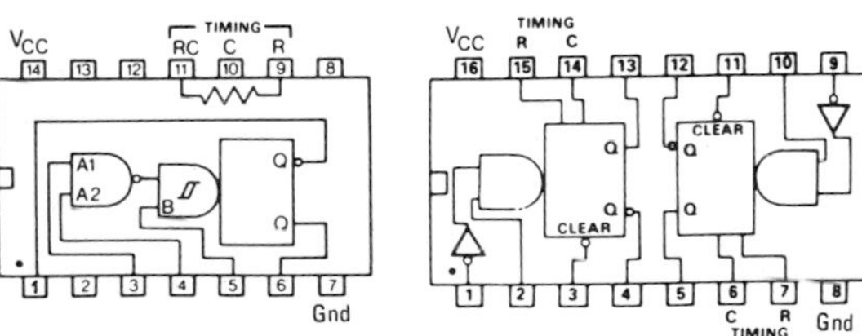

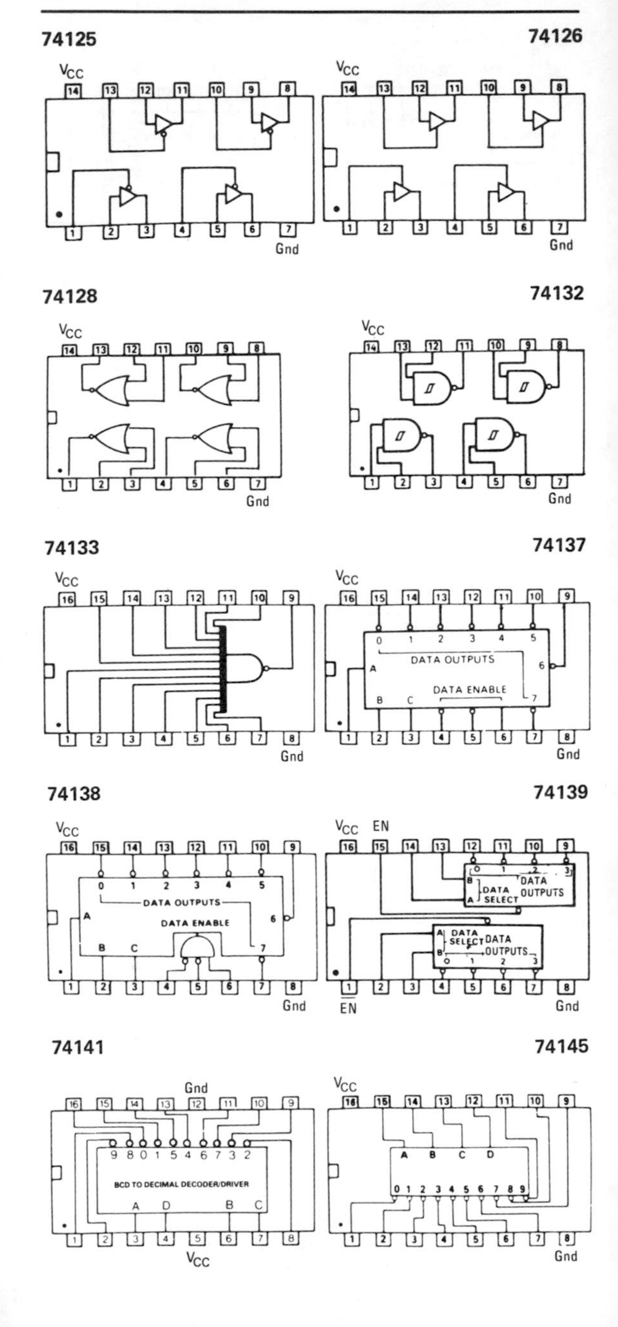
74125
74126
74128
74132
74133
74137
74138
74139
74141
74145
VCC
Gnd
DATA OUTPUTS
DATA ENABLE
DATA SELECT
EN
BCD TO DECIMAL DECODER/DRIVER

74147

74148

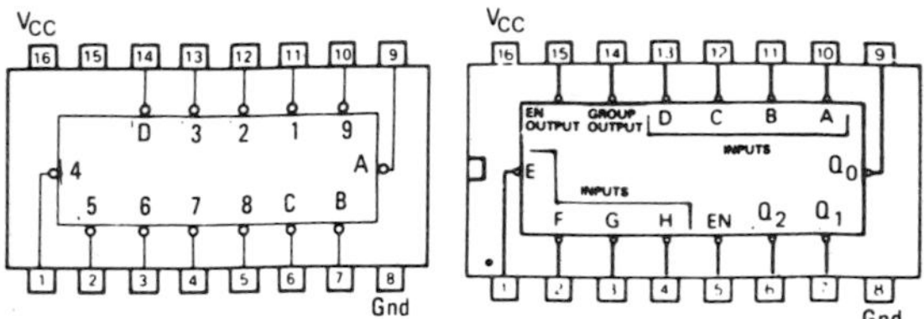

74151

74153

VCC
4 5 6 7 A B
DATA SELECT
3 DATA INPUTS C
2 1 0 Q Q̄ EN
Gnd
A SELECT
3 2 1 0 Q
DATA INPUTS
EN B B A A
EN DATA INPUTS B B A A
3 2 1 0 Q
B SELECT

74154

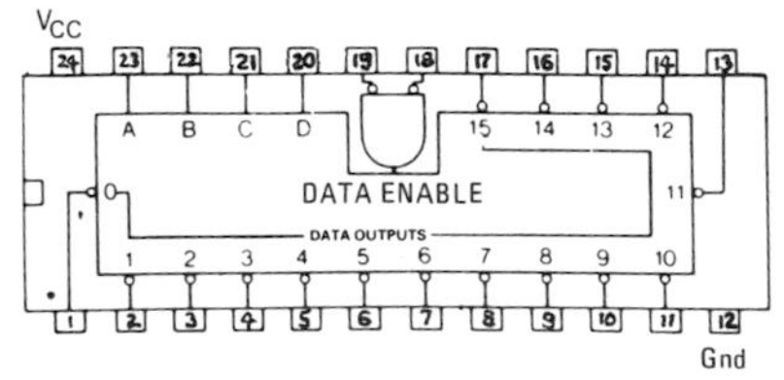

74155

74156

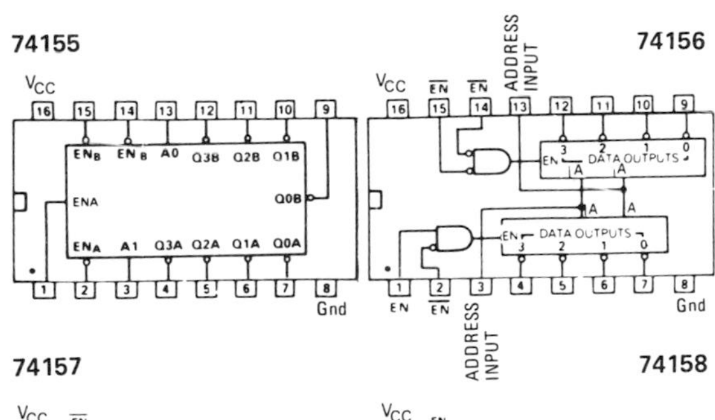

74157

74158

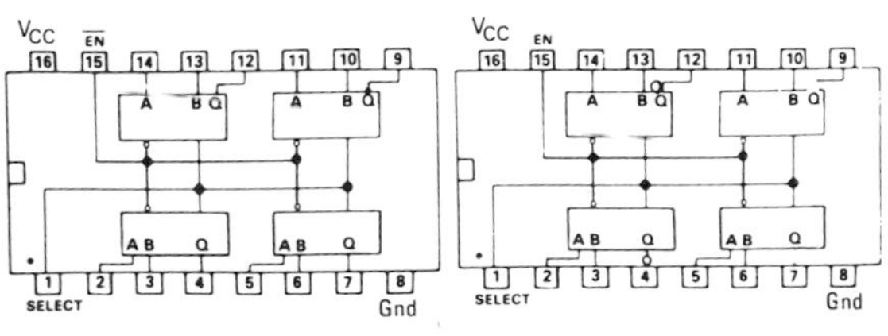

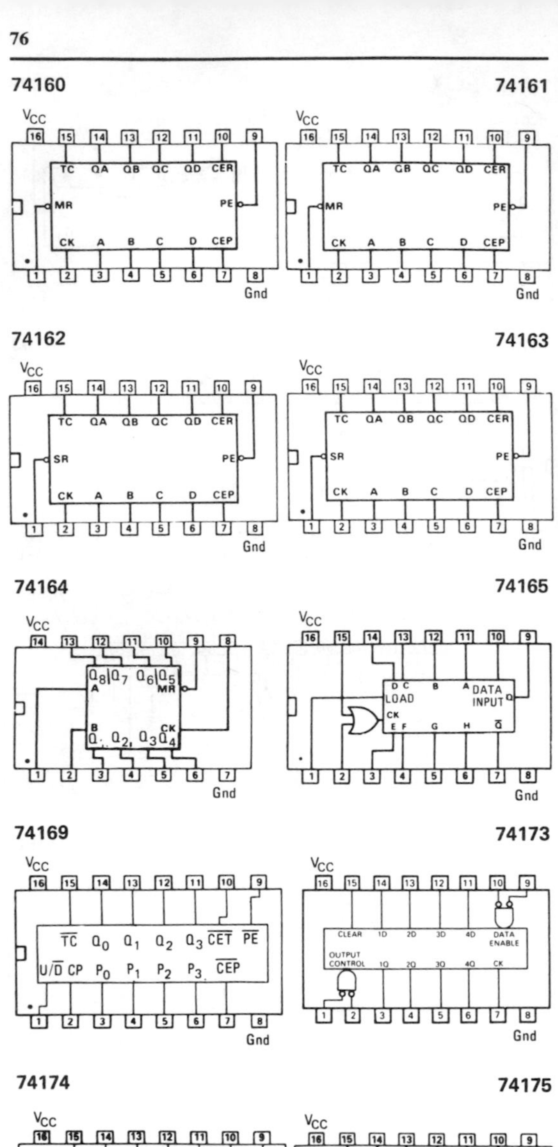
74160
74161
VCC
TC QA QB QC QD CER
MR PE
CK A B C D CEP
Gnd
74162
74163
TC QA QB QC QD CER
SR PE
CK A B C D CEP
74164
74165
Q8 Q7 Q6 Q5
A MR
B CK
Q1 Q2 Q3 Q4
D C B A DATA INPUT
LOAD
CK
E F G H
74169
74173
TC Q0 Q1 Q2 Q3 CET PE
U/D CP P0 P1 P2 P3 CEP
CLEAR 1D 2D 3D 4D DATA ENABLE
OUTPUT CONTROL 1Q 2Q 3Q 4Q CK

74174

74175

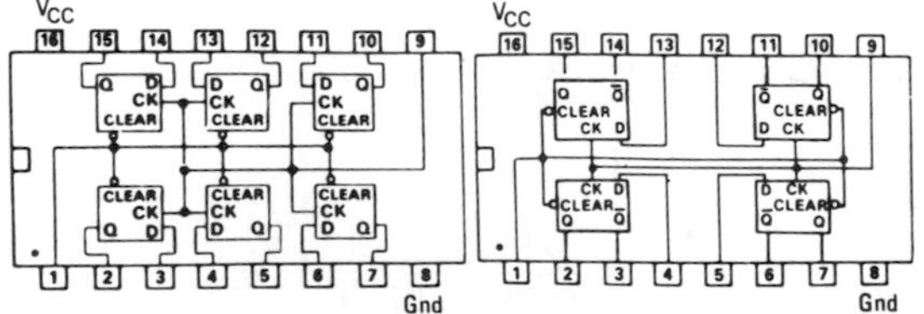
VCC
Q D
CK
CLEAR
Gnd

74180

74181

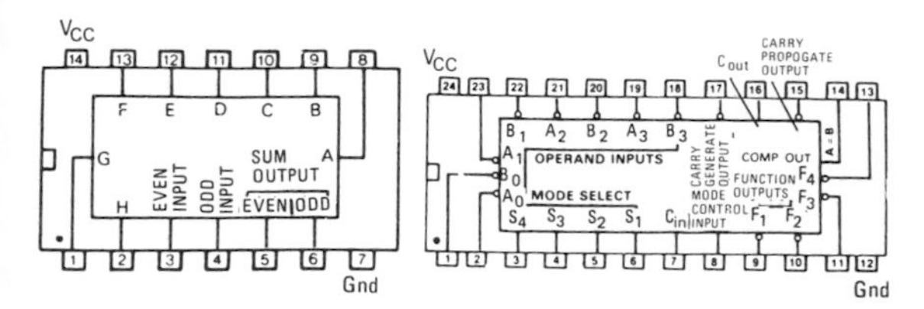

74191

74192

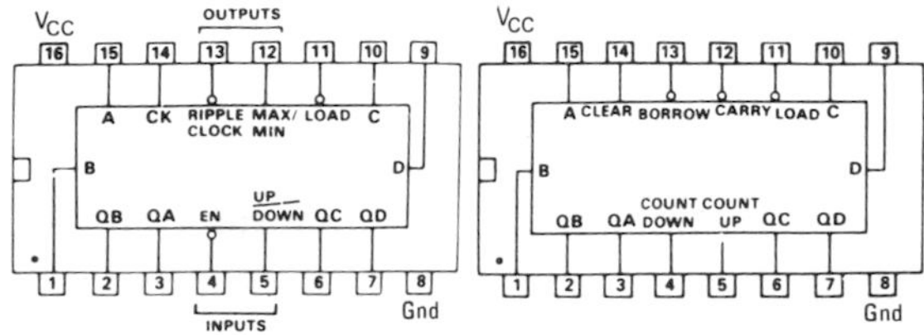

74193

74194

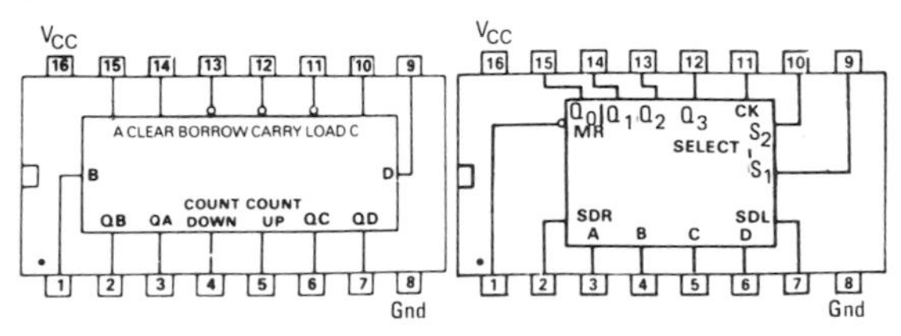

74195

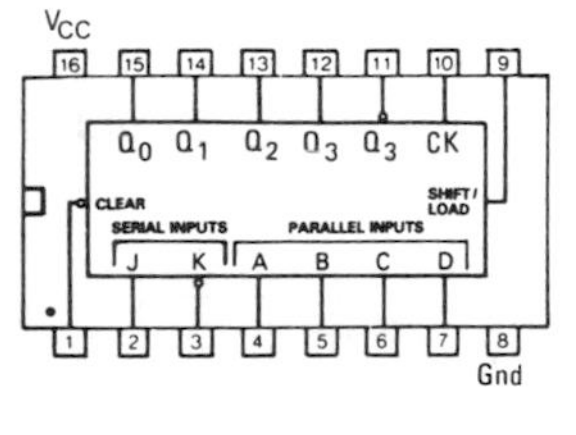

74196

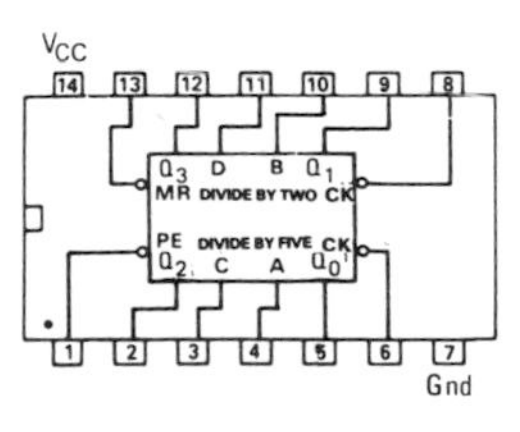

74197

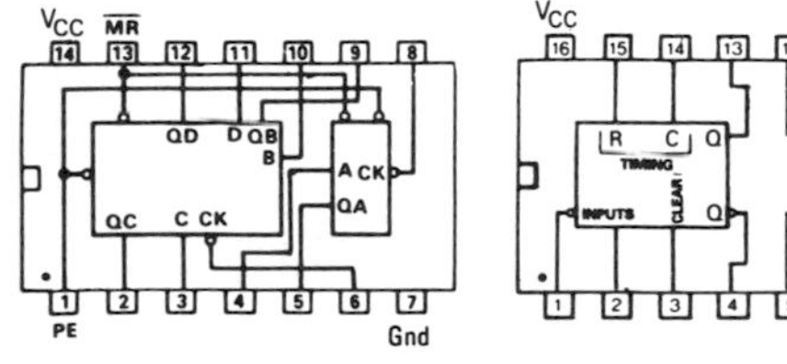

74221

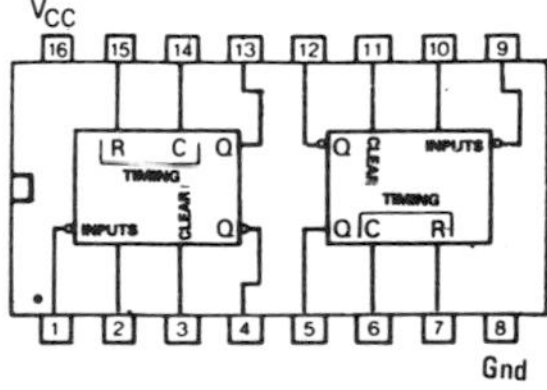

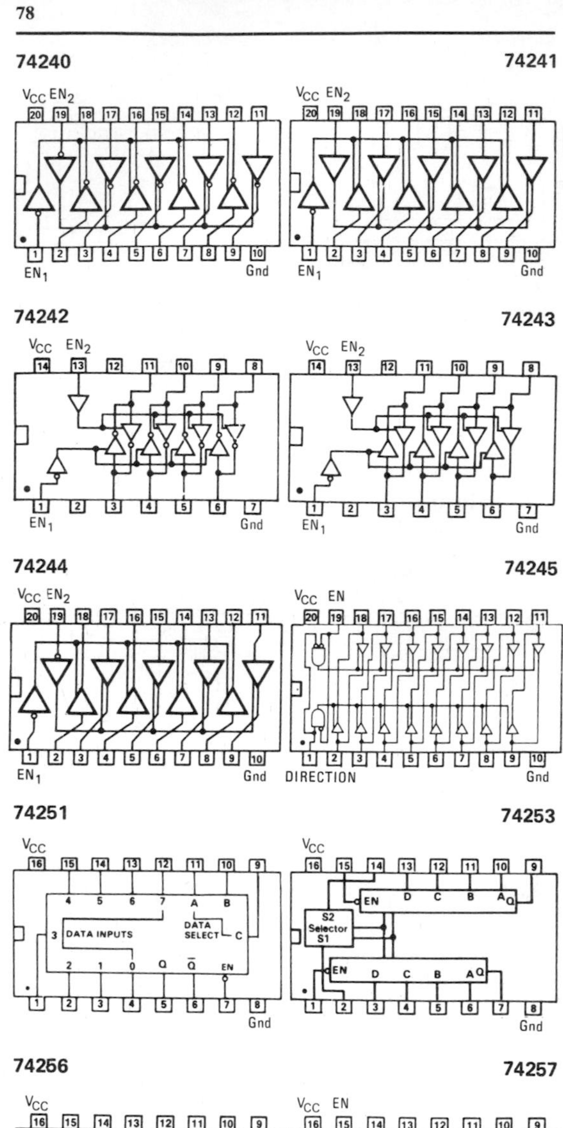
74240
74241
VCC EN2
EN1
Gnd
VCC EN2
EN1
Gnd
74242
74243
VCC EN2
EN1
Gnd
VCC EN2
EN1
Gnd
74244
74245
VCC EN2
EN1
Gnd
VCC EN
DIRECTION
Gnd
74251
74253
VCC
4 5 6 7 A B
DATA INPUTS
DATA SELECT
C
3
2 1 0 Q Q̄ EN
Gnd
VCC
EN D C B A Q
S2 Selector S1
EN D C B A Q
Gnd

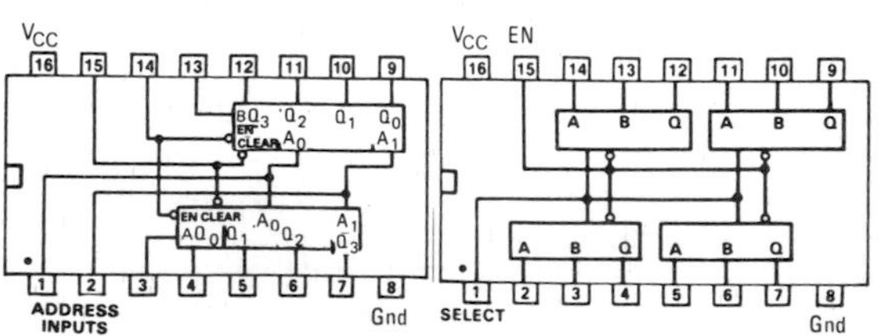
74256
74257
VCC
B Q3 Q2 Q1 Q0
EN CLEAR A0 A1
EN CLEAR A0 A1
A Q0 Q1 Q2 Q3
ADDRESS INPUTS
Gnd
VCC EN
A B Q
A B Q
A B Q
A B Q
SELECT
Gnd

74258

74259

74266

74273

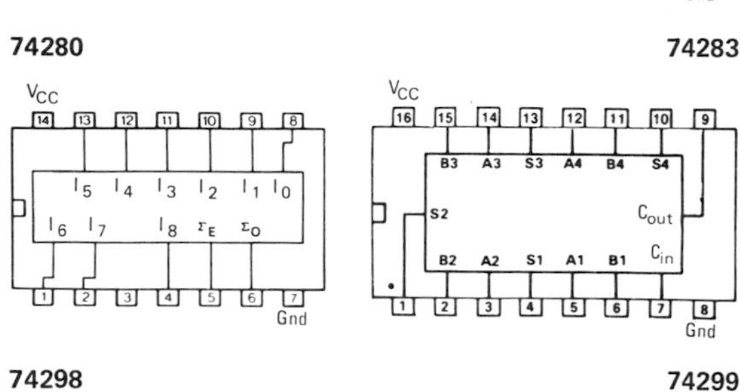

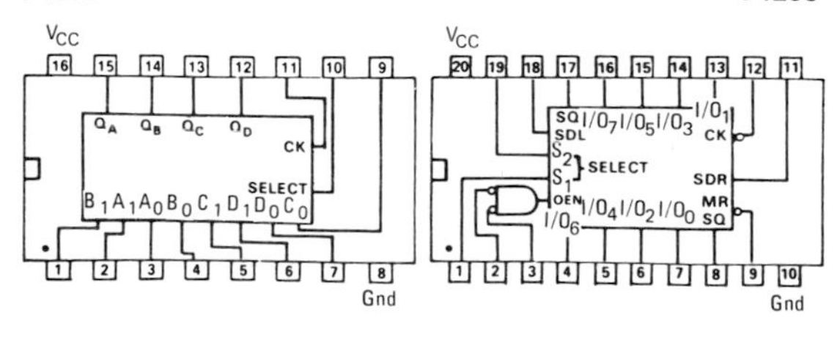

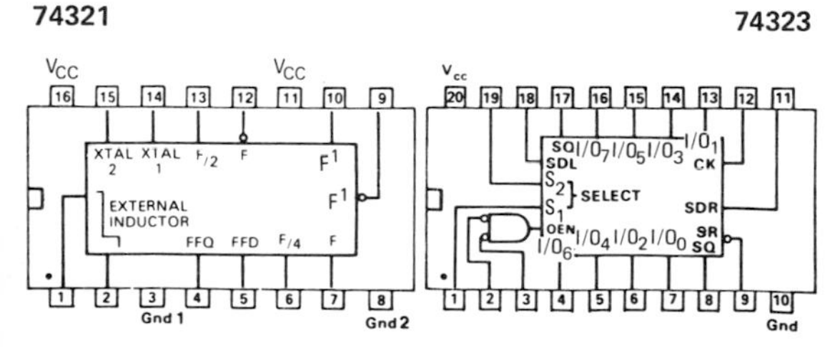

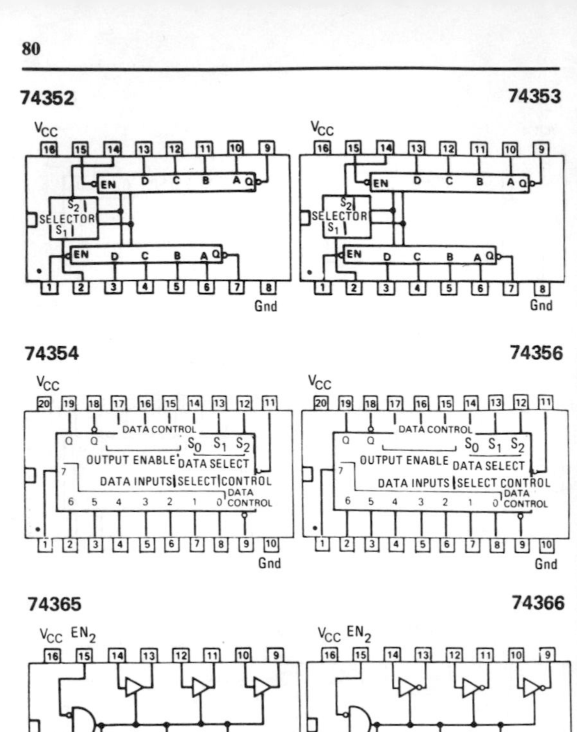
74352
74353
VCC
EN D C B A Q
S2
SELECTOR
S1
Gnd
74354
74356
DATA CONTROL
OUTPUT ENABLE
S0 S1 S2
DATA SELECT
DATA INPUTS
SELECT CONTROL
DATA CONTROL

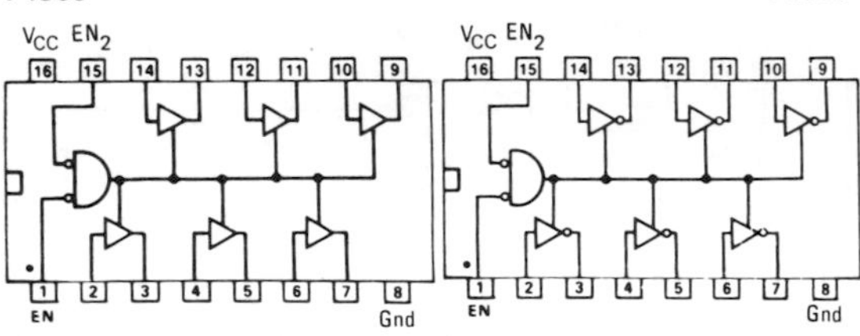
74365
74366
VCC EN2
EN
Gnd

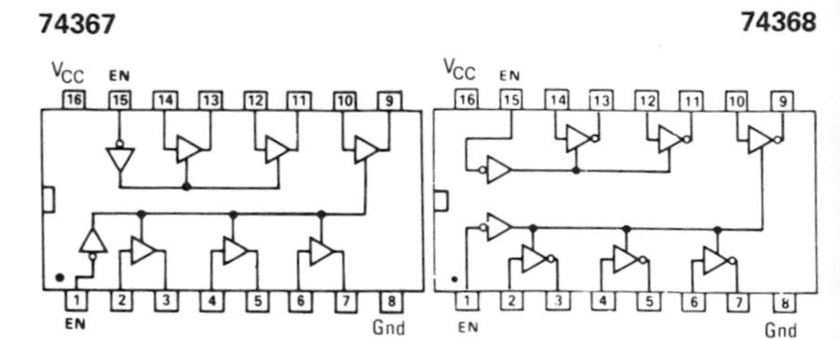
74367
74368
VCC EN
EN
Gnd

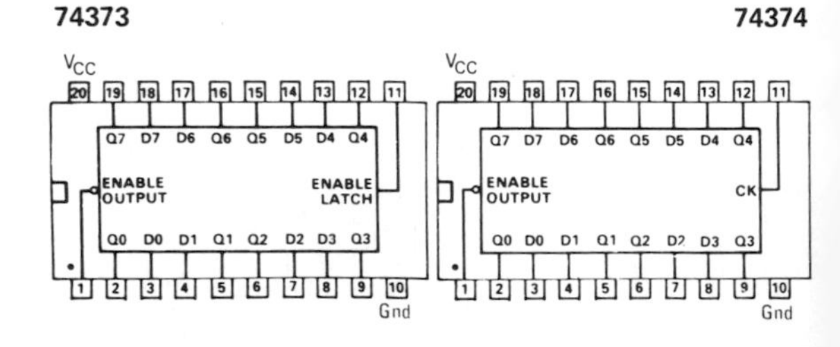
74373
74374
VCC
Q7 D7 D6 Q6 Q5 D5 D4 Q4
ENABLE OUTPUT
ENABLE LATCH
CK
Q0 D0 D1 Q1 Q2 D2 D3 Q3
Gnd

74377

V_{CC} CLOCK
20 19 18 17 16 15 14 13 12 11
D Q CK G
1 2 3 4 5 6 7 8 9 10
$\overline{EN}$ Gnd

74378

V_{CC}
16 15 14 13 12 11 10 9
Q5 F E Q4 D Q3
EN CK
Q0 A B Q1 C Q2
1 2 3 4 5 6 7 8
Gnd

74381

V_{CC}
20 19 18 17 16 15 14 13 12 11
A_2 B_2 A_3 B_3 C_n P G F_3 F_2
A_1 B_1 A_0 B_0 S_0 S_1 S_2 F_0 F_1
1 2 3 4 5 6 7 8 9 10
Gnd

74390

V_{CC}
16 15 14 13 12 11 10 9
MR Q_0 CK_1 Q_1 Q_2 Q_3 CK_0
CK_0 MR Q_0 CK_1 Q_1 Q_2 Q_3
1 2 3 4 5 6 7 8
Gnd

74393

V_{CC}
14 13 12 11 10 9 8
MR CK Q_0 Q_1 Q_2 Q_3
CK MR Q_0 Q_1 Q_2 Q_3
1 2 3 4 5 6 7
Gnd

74395

V_{CC}
16 15 14 13 12 11 10 9
Q_A Q_B Q_C Q_D CK
CLEAR CASCADE OUTPUT OUTPUT CONTROL
SERIAL INPUT A B C D LOAD SHIFT
1 2 3 4 5 6 7 8
Gnd

74399

74423

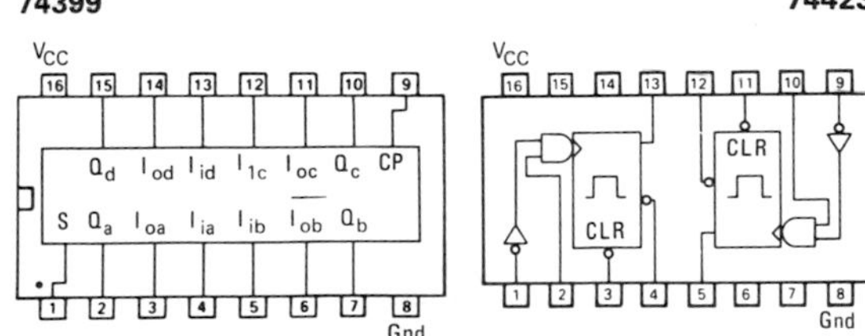

74442, 443, 444

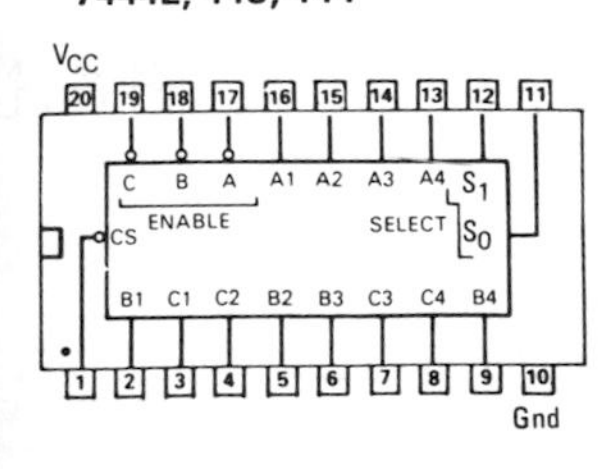

74533

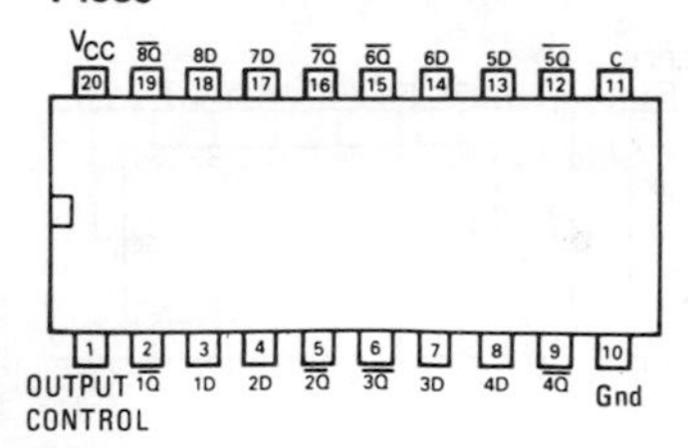

74534

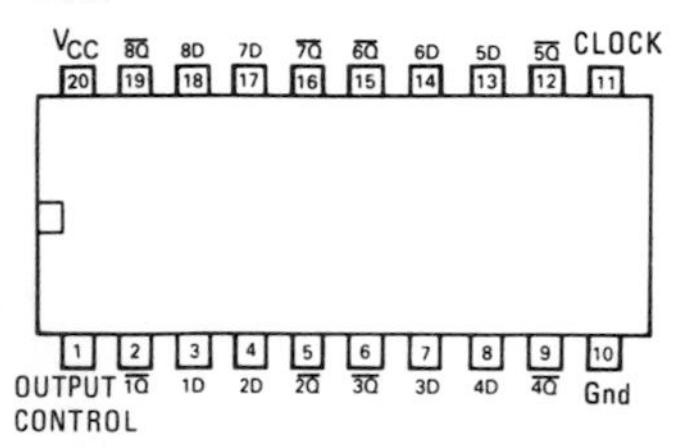

74563

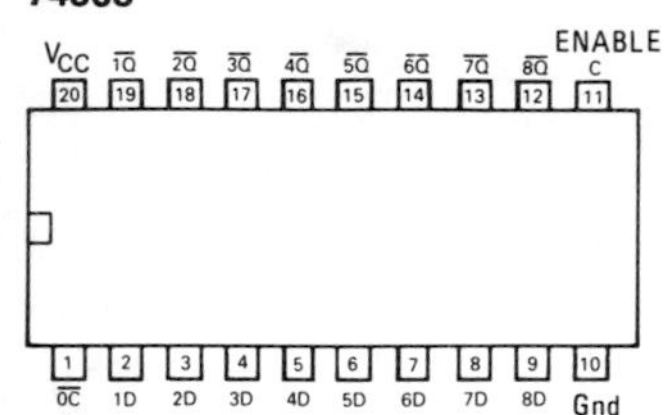

74564

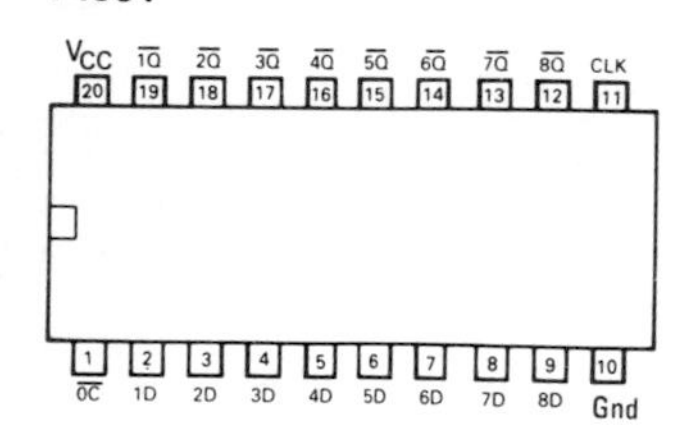

74573

74574

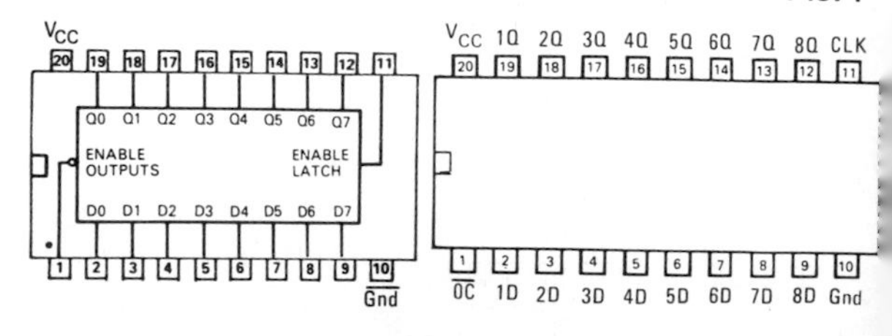

74580

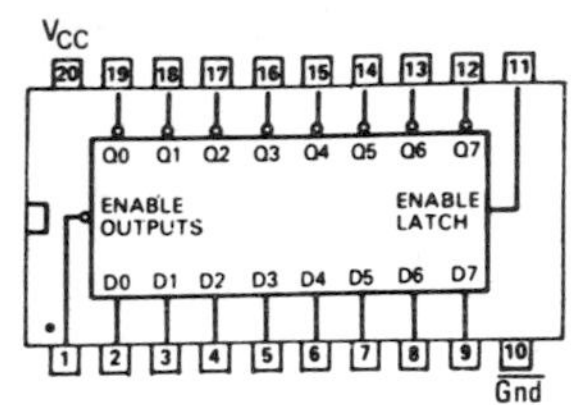

74620

74625

74640

74643

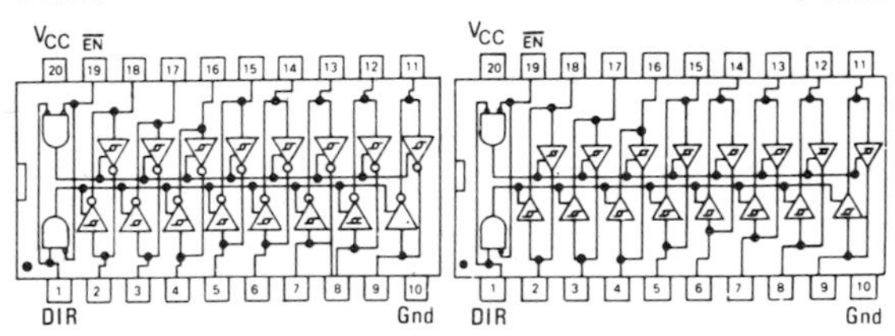

74655

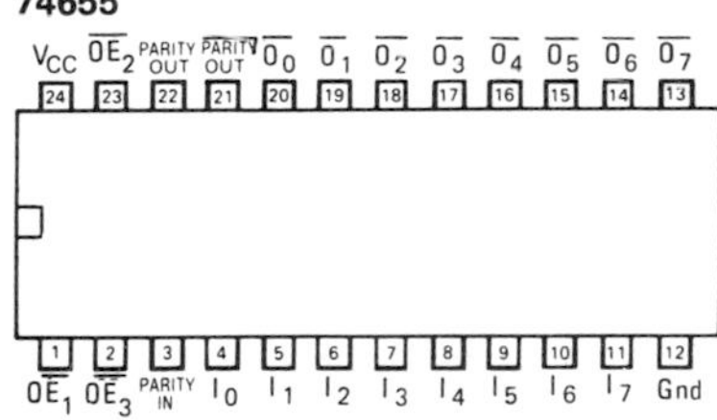

74656

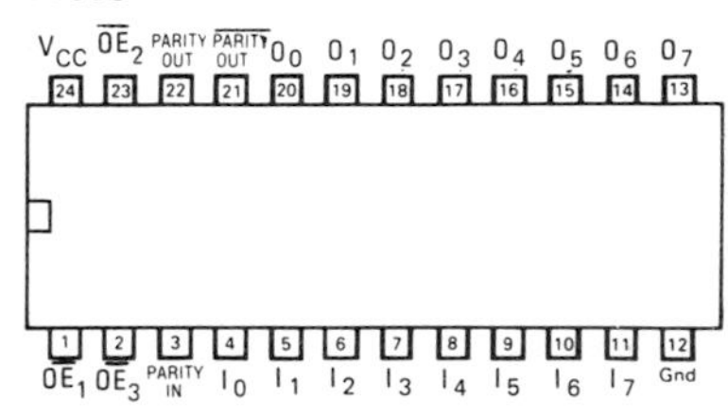

74657

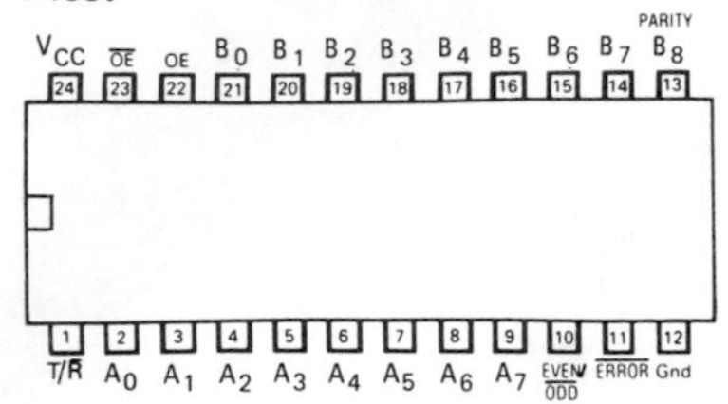

74669

74670

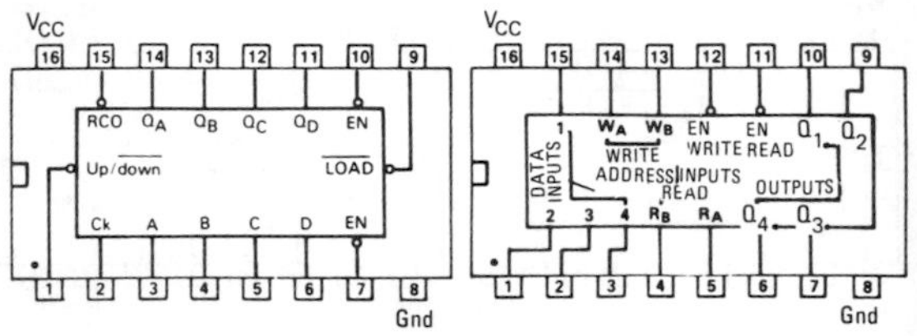

74673

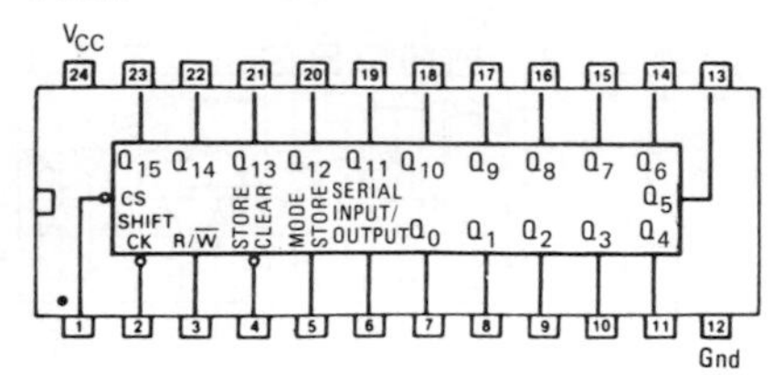

74674

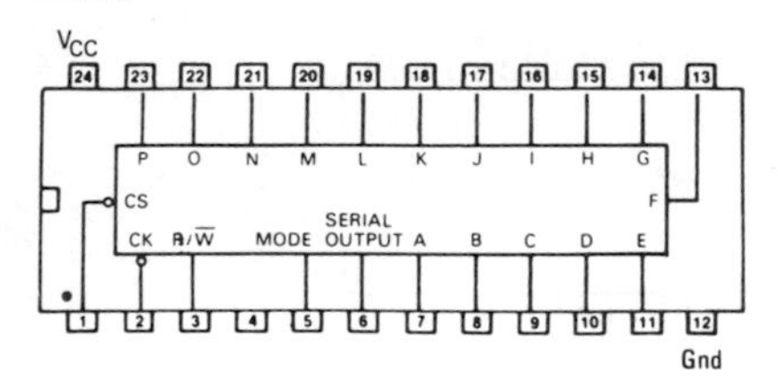

74682

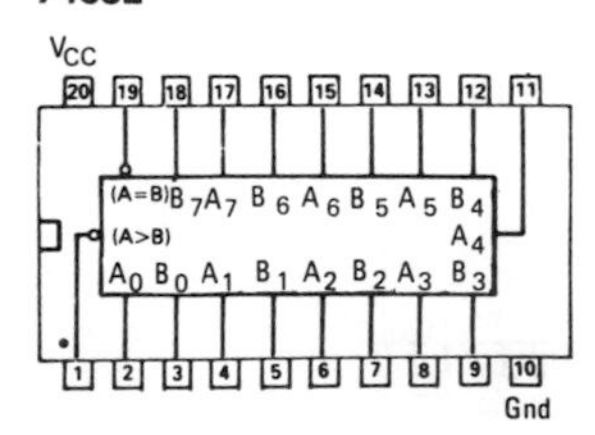

74688

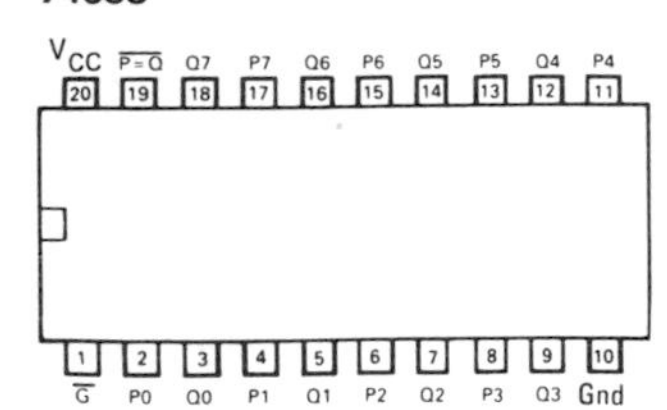

741242

741243

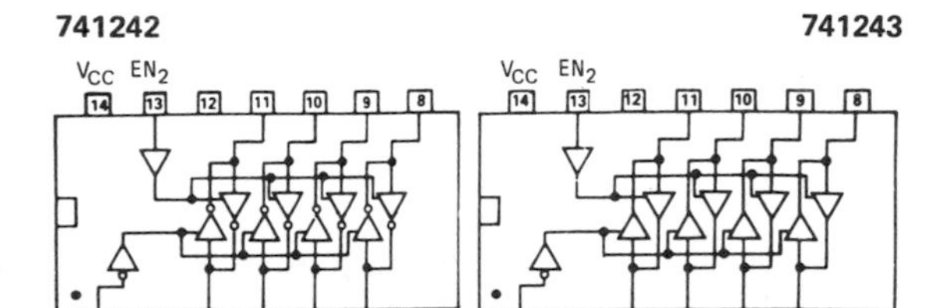

744002

744017

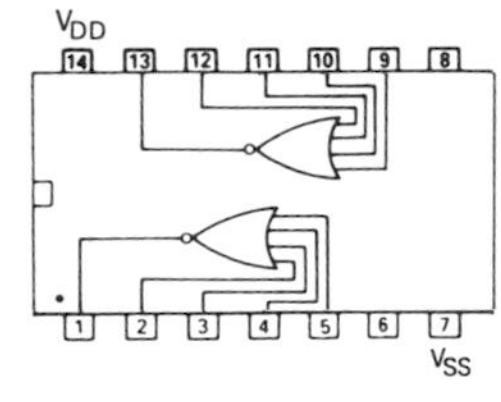

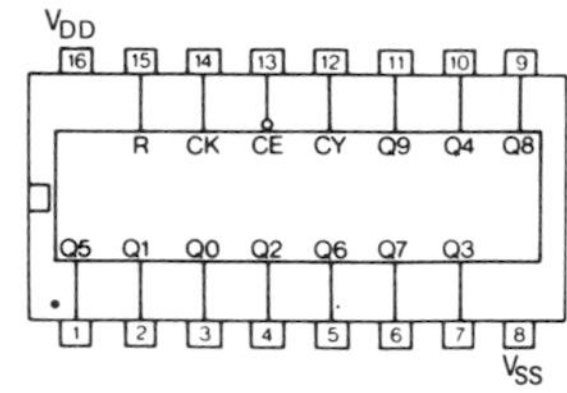

744020

744040

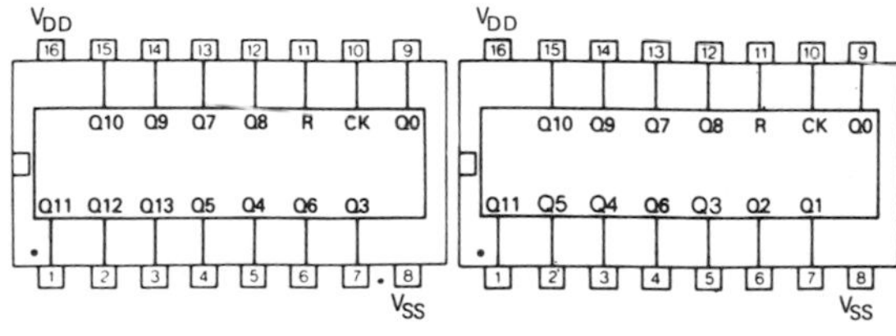

744049

744050

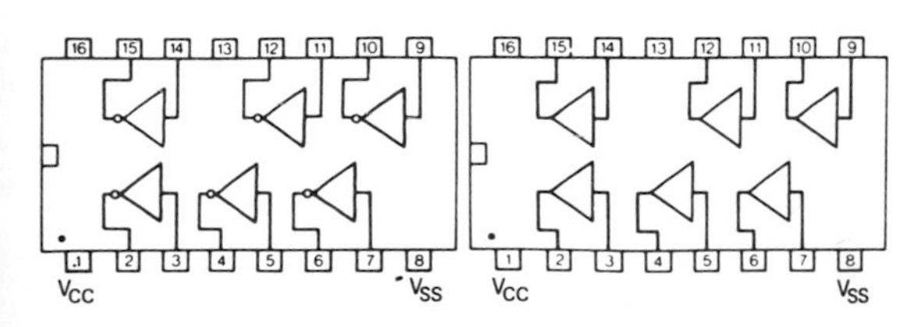

744060

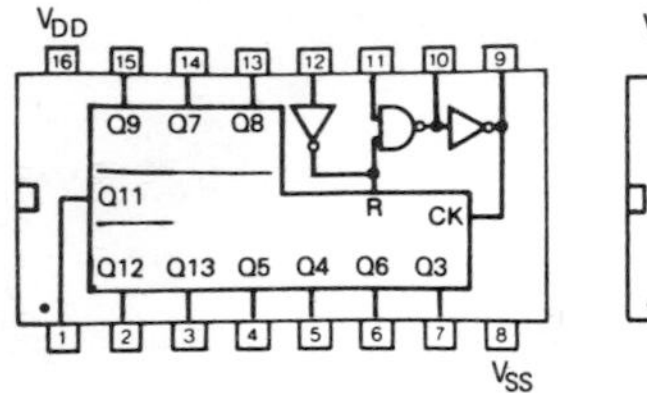

744075

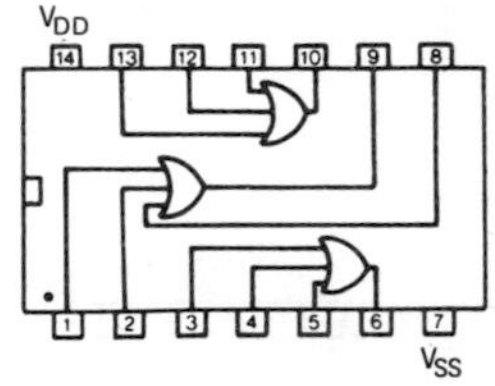

744078

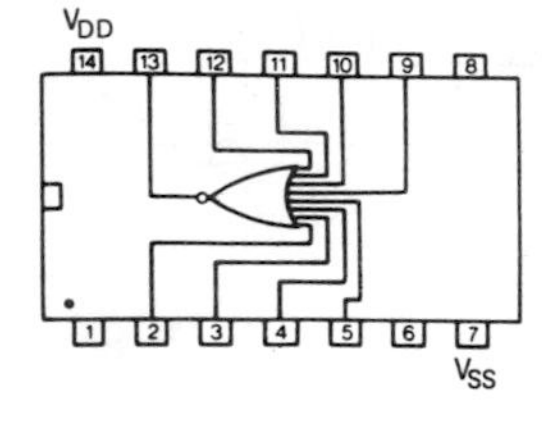

744511

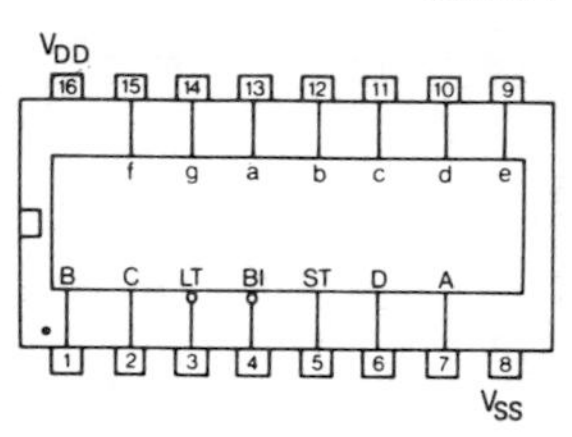

744514

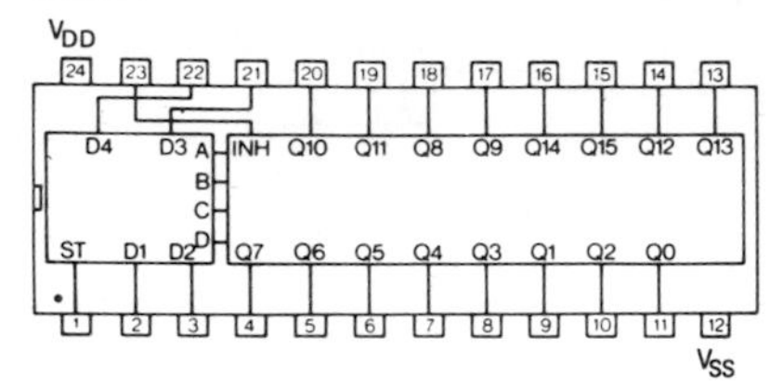

744538 **744543**

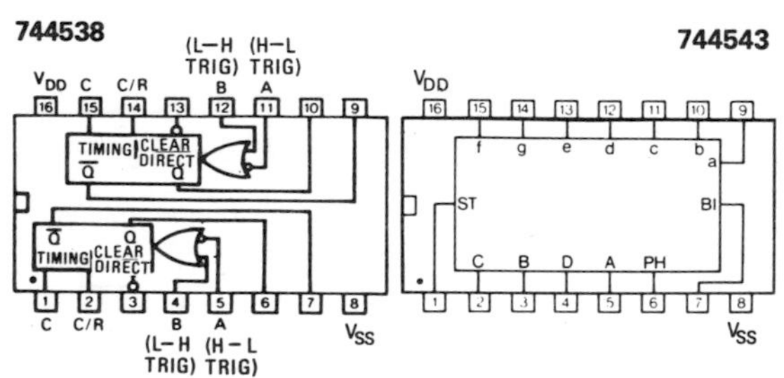

CMOS data

Selection by device number

Device	Description
4000	Dual 3-input NOR gate plus Inverter
4001	Quad 2-input NOR Gate

4002	Dual 4-input NOR Gate
4006	18-Stage Static Shift Register
4007	Dual Complementary Pair Plus Inverter
4008	4-Bit full Adder with Parallel Carry
4009	Hex Buffer/Converter (Inverting)
4010	Hex Buffer/Converter (Non-Inverting)
4011	Quad 2-Input NAND Gate
4012	Dual 4-Input NAND Gate
4013	Dual 'D' Flip-Flop with Set/Reset
4014	8-Stage Static Shift Register
4015	Dual 4-Stage Static Shift Register
4016	Quad Bilateral Switch
4017	Docade Counter/Divider
4018	Presettable Divide-By-'N' Countor
4019	Quad AND-OR Select Gate
4020	14-Stage Binary Ripple Counter
4021	8-Stage Static Shift Register
4022	Divide-by-8 Counter/Divider
4023	Triple 3-Input NAND Gate
4024	7-Stage Binary Counter
4025	Triple 3-Input NOR Gate
4026	Decade Counter/Divider
4027	Dual J-K Master Slave Flip-Flop
4028	BCD-to-Decimal Decoder
4029	Presettable Up/Down Counter
4030	Quad Exclusive-OR Gate
4032	Triple Serial Adder
4035	4-Stage Parallel IN/OUT Shift Register
4038	Triple Serial Adder
4040	12/Stage Binary Ripple Counter
4042	Quad Clocked 'D' Latch
4043	Quad, 3-state R-S Latch
4044	Quad, 3-state R-S Latch
4046	Micropower Phase-Locked Loop
4047	Multivibrator, Astable/Monostable
4049	Hex Buffer/Converter (Inverting)
4050	Hex Buffer/Converter (Non-Inverting)
4051	Single 8-Channel Multiplexer
4052	Differential 4-Channel Multiplexer
4053	Triple, 2-input Analogue Multiplexer
4054	4-Line Liquid Crystal Display Driver
4056	BCD-7-Segment Decoder/Driver
4059	Programmable Divide-by-N Counter
4060	14-Stage Counter and Oscillator
4061	256-Word X 1-Bit Static Ram
4066	Quad Bilateral Switch
4068	8-Input NAND Gate
4069	Hex Inverter
4070	Quad Exclusive OR Gate
4071	Quad 2-Input OR Gate
4072	Dual, 4-input OR Gate
4073	Triple, 3-input AND Gate
4075	Triple, 3-input OR Gate
4076	Quad, 3-state D Register
4077	Quad Exclusive NOR Gate
4078	8-input NOR Gate
4081	Quad 2-Input AND Gate
4082	Dual 4-Input AND Gate
4085	Dual 2-Wide 2-Input AOI Gate

4086	Expendable 4-Wide 2-Input AOI Gate
4093	Quad 2-Input NAND Schmitt Trigger
4094	8-stage Shift Register, with Storage
4099	8-Bit Addressable Latch
40106	Hex, Inverting Schmitt Buffers
4160	Asynchronous Decade Counter with Clear
4161	Asynchronous 4-bit Binary Counter with Clear
4162	Synchronous Decade Counter with Clear
4163	Synchronous 4-bit Binary Counter with Clear
4502	Strobed Hex Inverting Buffer
4508	Dual 4-bit Latch
4510	BCD UP/DOWN Counter
4511	BCD-to-Segment Decoder/Driver
4512	8-channel Data Selector
4513	BCD-to-7-segment Latch/Driver
4514	1 to 16 Decoder (Output High)
4515	1 to 16 Decoder (Output Low)
4516	Binary UP/DOWN Counter
4518	Dual BCD UP Counter
4519	Quad, 2-input Multiplexer
4520	Dual 4-bit Binary Counter
4521	24-stage Frequency Divider
4522	BCD Programmable Divider
4526	Binary Programmable Divider
4527	BCD Rate Multiplier
4528	Dual Retriggerable Monostable
4529	Dual 4-channel Analogue Selector
4530	Dual 5-bit Majority Gate
4531	12-bit Parity Tree
4532	8-bit Priority Encoder
4536	Programmable Timer
4538	Dual Monostable Multivibrator
4539	Dual 4-bit Multiplexer
4541	Programmable Timer
4543	BCD-to-7-segment Latched LCD Driver
4551	Quad 2-input Analogue Multiplexer
4553	3-digit BCD Counter
4554	2 × 2 Binary Multiplier
4556	Dual Binary to 1-of-4 Decoder
4560	BCD Adder
4561	9's Complementer
4566	Timebase Generator
4580	4 × 4 Multiport Register
4581	4-bit Arithmetic Logic Unit
4582	Carry Look Ahead 4-bit Magnitude
4583	Dual Schmitt Gates
4585	4-bit Magnitude Comparator
4597	8-bit 3-state Bus Latch
4598	8-bit 3-state Bus Latch
4599	8-bit Addressable Latch
45100	4 × 4 Crosspoint Switch

Selection by function

Gates

AND

Triple 3-input	**4073**
Quad 2-input	**4081**

OR

Quad 2-input	**4071**

Dual 4-input	**4072**
Triple 3-input	**4075**
Exclusive OR	
Quad 2-input	**4070**
NAND	
Quad 2-input	**4011**
Dual 4-input	**4012**
Triple 3-input	**4023**
8-input	**4068**
NOR	
Quad 2-input	**4001**
Dual 4-input	**4002**
Triple 3-input	**4025**
8-input	**4078**
Exclusive NOR	
Quad 2-input	**4077**
Schmitt	
Quad 2-input NAND	**4093**
Hex inverting	**40106**
Dual	**4583**
Majority	
Dual 5-bit	**4530**
Buffers	
Hex inverting	**4049**
Hex	**4050**
Hex inverting	**4069**
Strobed Hex inverting	**4502**

Flip-flops (bistables)

Dual D-type	**4013**
Dual J-K	**4027**
Quad latch	**4042**
Quad R-S latch 3-state	**4043**
Quad R-S latch 3-state	**4044**
Quad D register 3-state	**4076**
8-bit addressable latch	**4099**
Dual 4-bit latch	**4508**
4 × 4 multiport register	**4580**
8-bit bus latch 3-state	**4597**
8-bit bus latch 3-state	**4598**
8-bit addressable latch	**4599**

Counters

decade/divider	**4017**
divide by n	**4018**
14-bit binary	**4020**
Octal/divider	**4022**
7-stage binary	**4024**
Presettable binary/BCD, up/down	**4029**
12-bit binary	**4040**
14-bit binary	**4060**
Decade async. clear	**4160**
4-bit binary async. clear	**4161**
Decade sync. clear	**4162**
4-bit binary sync. clear	**4163**
BCD up/down	**4510**
Binary up/down	**4516**
Dual BCD up	**4518**
Dual 4-bit binary	**4520**
24-stage frequency divider	**4521**

BCD programmable divider	**4522**
Binary programmable divider	**4526**
3-digit BCD	**4553**
Shift registers	
8-bit	**4014**
Dual 4-bit	**4015**
8-bit	**4021**
4-bit FIFO	**4035**
8-stage with storage	**4094**
Encoders, decoders/drivers	
Decoders	
BCD-decimal, binary-octal	**4028**
BCD-7-segment latch/driver	**4511**
BCD-7-segment latch/driver	**4513**
4-bit latch, 4-to-16 line	**4514**
4-bit latch, 4-to-16 line inverted outputs	**4515**
BCD-7-segment latched LCD driver	**4543**
Dual binary to 1-of-4	**4556**
Encoders/multiplexers	
8-input analogue multiplexer	**4051**
Dual 4-input analogue multiplexer	**4052**
Triple 2-input analogue multiplexer	**4053**
8-channel data selector	**4512**
Quad 2-input multiplexer	**4519**
Dual 4-channel analogue selector	**4529**
8-bit priority encoder	**4532**
Dual 4-input multiplexer	**4539**
Quad 2-input analogue multiplexer	**4551**
Arithmetic functions	
4-bit full adder	**4008**
Triple serial adder + logic	**4032**
Triple serial adder − logic	**4038**
BCD rate multiplier	**4527**
12-bit parity tree	**4531**
2 × 2 binary multiplier	**4554**
BCD adder	**4560**
9's complementer	**4561**
4-bit arithmetic logic unit	**4581**
Carry look ahead	**4582**
4-bit magnitude comparator	**4585**
Miscellaneous	
Quad switch	**4016**
Phase locked loop	**4046**
Mono/astable multivibrator	**4047**
Quad switch	**4066**
Dual resettable monostable	**4528**
Programmable timer	**4536**
Dual monostable multivibrator	**4538**
Programmable timer	**4541**
Industrial timebase generator	**4566**
4 × 4 crosspoint switch	**45100**

CMOS pinouts

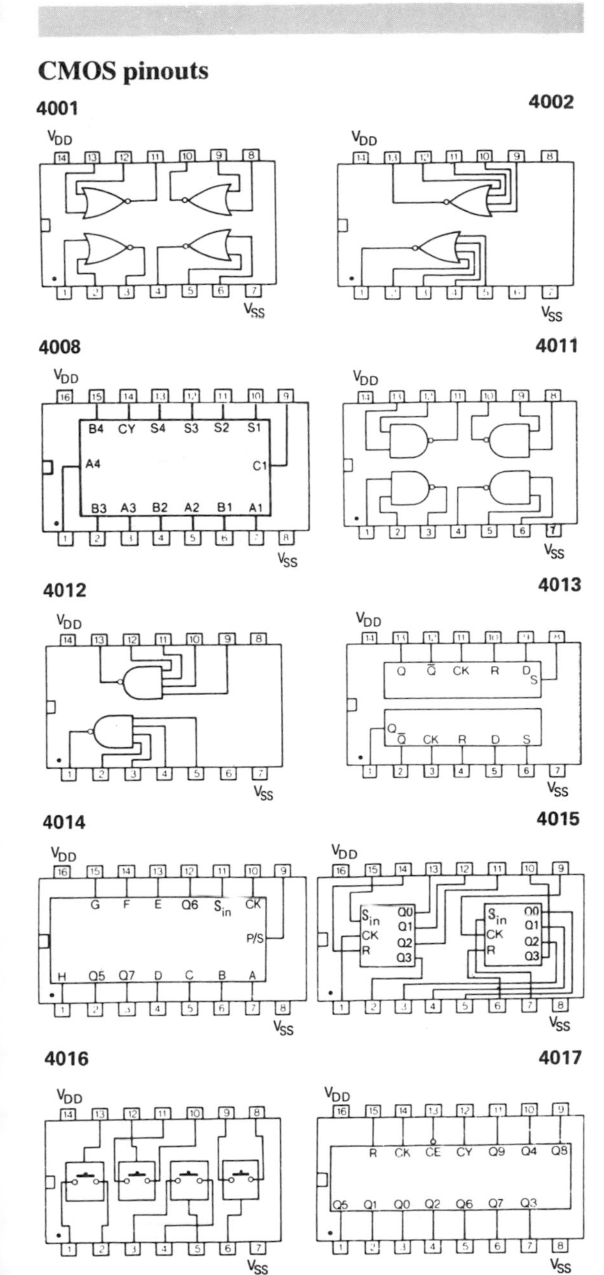

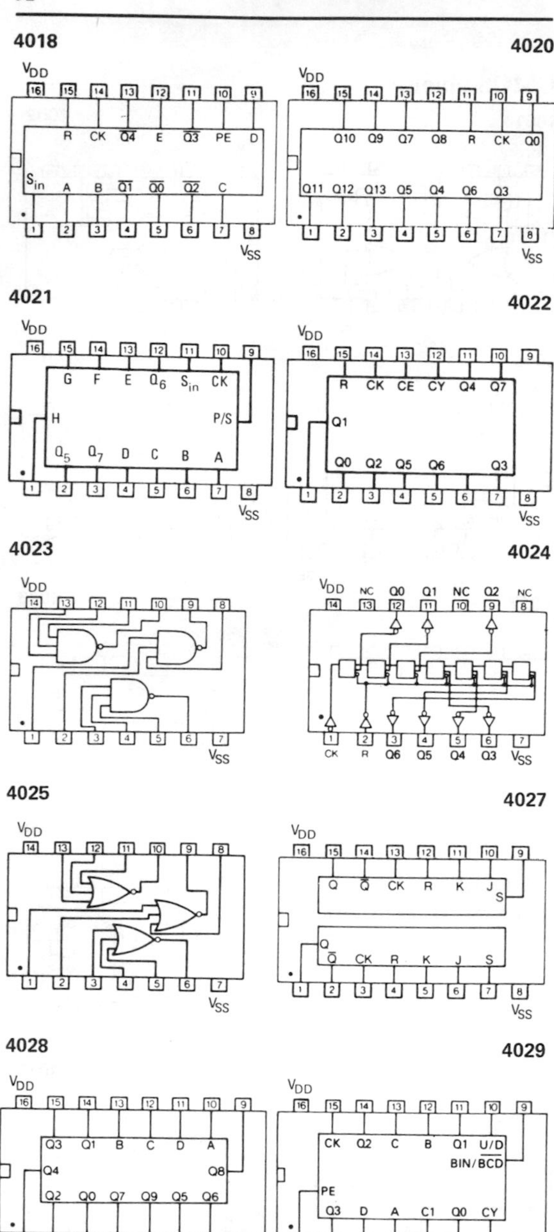
4018
VDD
16 15 14 13 12 11 10 9
R CK Q4 E Q3 PE D
Sin A B Q1 Q0 Q2 C
1 2 3 4 5 6 7 8
VSS
4020
VDD
16 15 14 13 12 11 10 9
Q10 Q9 Q7 Q8 R CK Q0
Q11 Q12 Q13 Q5 Q4 Q6 Q3
1 2 3 4 5 6 7 8
VSS
4021
VDD
16 15 14 13 12 11 10 9
G F E Q6 Sin CK
H
P/S
Q5 Q7 D C B A
1 2 3 4 5 6 7 8
VSS
4022
VDD
16 15 14 13 12 11 10 9
R CK CE CY Q4 Q7
Q1
Q0 Q2 Q5 Q6 Q3
1 2 3 4 5 6 7 8
VSS
4023
VDD
14 13 12 11 10 9 8
1 2 3 4 5 6 7
VSS
4024
VDD
NC Q0 Q1 NC Q2 NC
14 13 12 11 10 9 8
1 2 3 4 5 6 7
CK R Q6 Q5 Q4 Q3 VSS
4025
VDD
14 13 12 11 10 9 8
1 2 3 4 5 6 7
VSS
4027
VDD
16 15 14 13 12 11 10 9
Q Q CK R K J
S
Q
Q CK R K J S
1 2 3 4 5 6 7 8
VSS
4028
VDD
16 15 14 13 12 11 10 9
Q3 Q1 B C D A
Q4
Q8
Q2 Q0 Q7 Q9 Q5 Q6
1 2 3 4 5 6 7 8
VSS
4029
VDD
16 15 14 13 12 11 10 9
CK Q2 C B Q1 U/D
BIN/BCD
PE
Q3 D A C1 Q0 CY
1 2 3 4 5 6 7 8
VSS

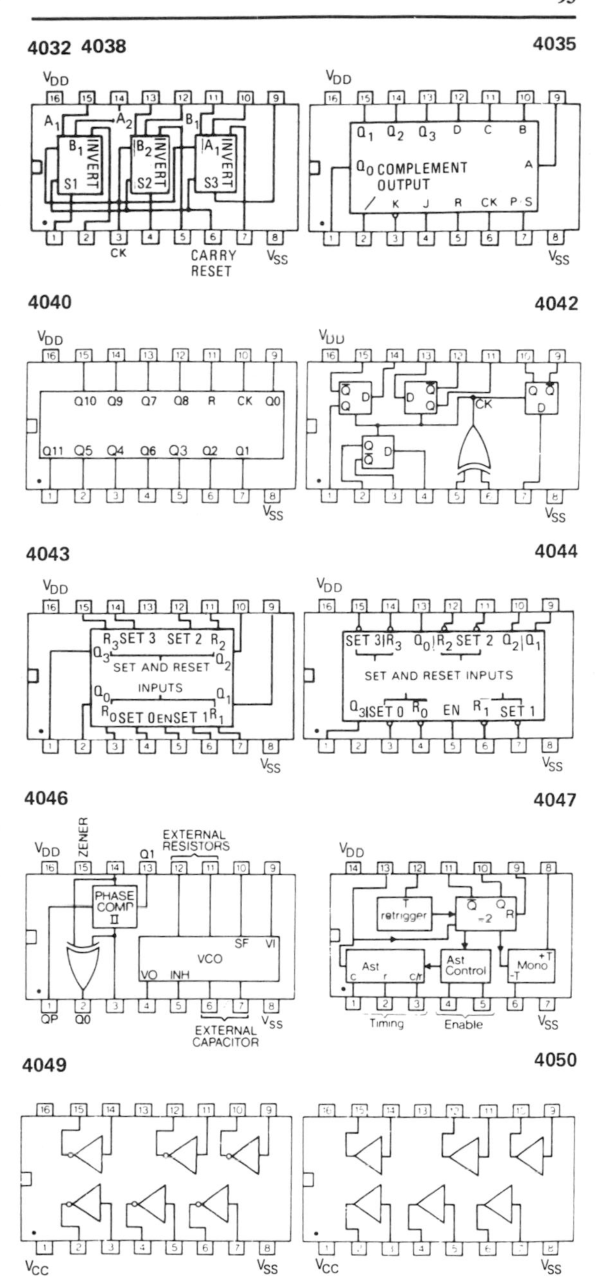
4032 4038
VDD
A1
A2
B1
INVERT
S1
S2
S3
CK
CARRY RESET
VSS
4035
Q1 Q2 Q3 D C B
Q0 COMPLEMENT OUTPUT
A
K J R CK P/S
4040
Q10 Q9 Q7 Q8 R CK Q0
Q11 Q5 Q4 Q6 Q3 Q2 Q1
4042
CK
4043
R3 SET 3 SET 2 R2
SET AND RESET INPUTS
R0 SET 0 EN SET 1 R1
4044
SET 3 R3 Q0 R2 SET 2 Q2 Q1
SET AND RESET INPUTS
Q3 SET 0 R0 EN R1 SET 1
4046
ZENER
Q1
EXTERNAL RESISTORS
PHASE COMP II
VCO
SF VI
VO INH
QP Q0
EXTERNAL CAPACITOR
4047
retrigger
÷2
Ast
Ast Control
Mono
Timing
Enable
4049
VCC
4050

4051

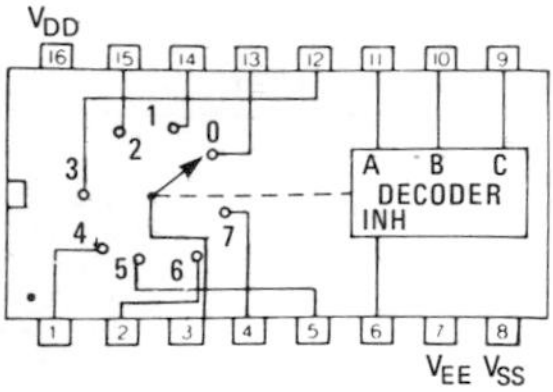

V_{DD}
A B C DECODER INH
V_{EE} V_{SS}

4052

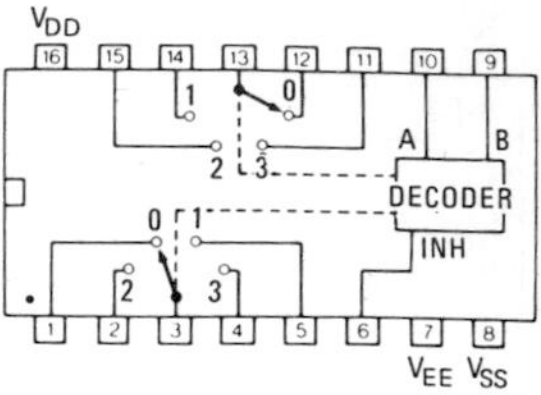

V_{DD}
A B DECODER INH
V_{EE} V_{SS}

4053

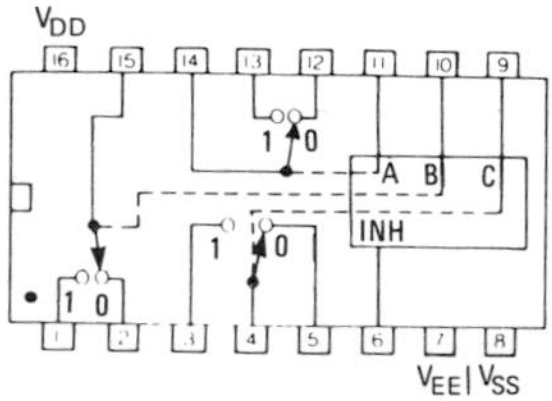

V_{DD}
A B C INH
V_{EE} V_{SS}

4060

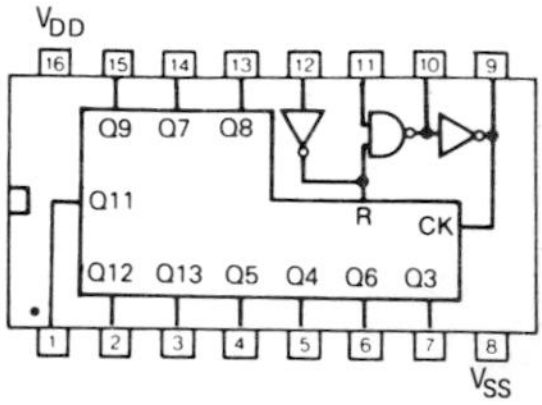

V_{DD}
Q9 Q7 Q8 Q11 R CK Q12 Q13 Q5 Q4 Q6 Q3
V_{SS}

4066

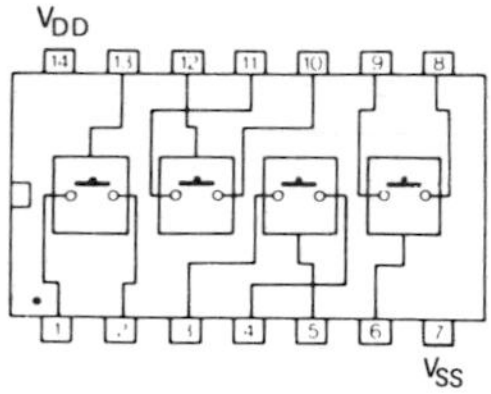

V_{DD}
V_{SS}

4068

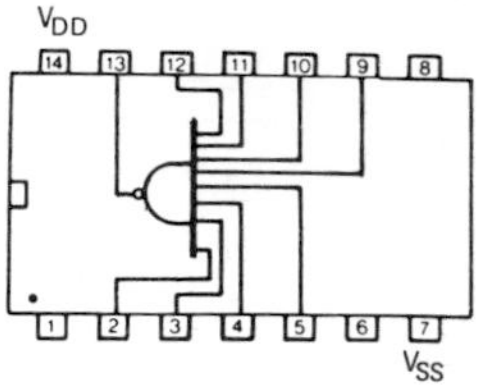

V_{DD}
V_{SS}

4069

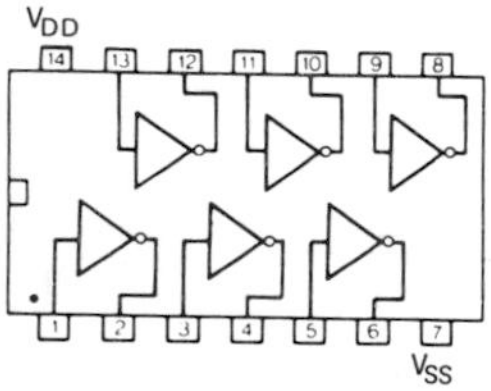

V_{DD}
V_{SS}

4070

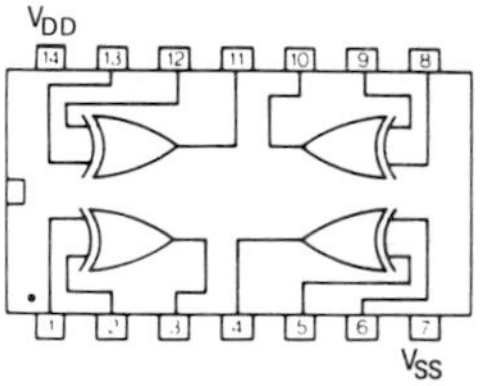

V_{DD}
V_{SS}

4071

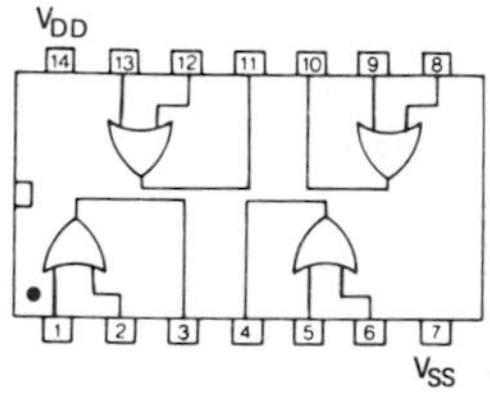

V_{DD}
V_{SS}

4072

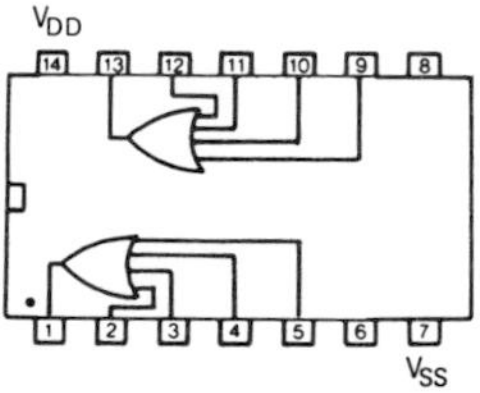

V_{DD}
V_{SS}

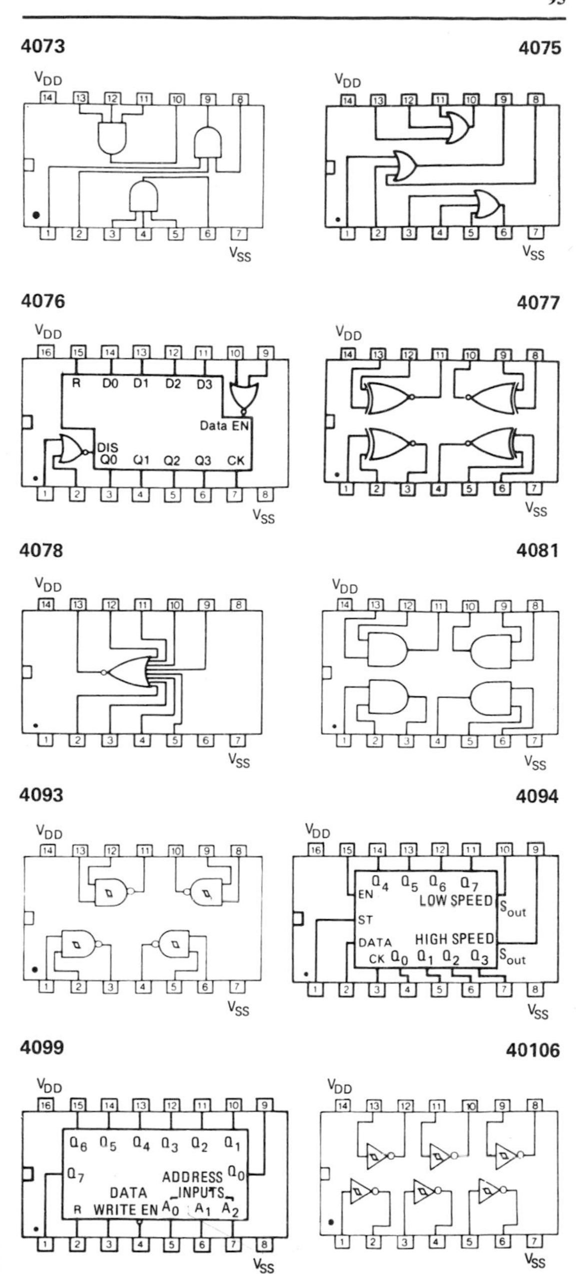
4073
V_{DD}
V_{SS}
4075
V_{DD}
V_{SS}
4076
V_{DD}
R D0 D1 D2 D3
Data EN
DIS
Q0 Q1 Q2 Q3 CK
V_{SS}
4077
V_{DD}
V_{SS}
4078
V_{DD}
V_{SS}
4081
V_{DD}
V_{SS}
4093
V_{DD}
V_{SS}
4094
V_{DD}
EN Q_4 Q_5 Q_6 Q_7
LOW SPEED S_{out}
ST
DATA HIGH SPEED
CK Q_0 Q_1 Q_2 Q_3 S_{out}
V_{SS}
4099
V_{DD}
Q_6 Q_5 Q_4 Q_3 Q_2 Q_1
Q_7 ADDRESS Q_0
DATA INPUTS
R WRITE EN A_0 A_1 A_2
V_{SS}
40106
V_{DD}
V_{SS}

4160 4161 4162 4163

4502

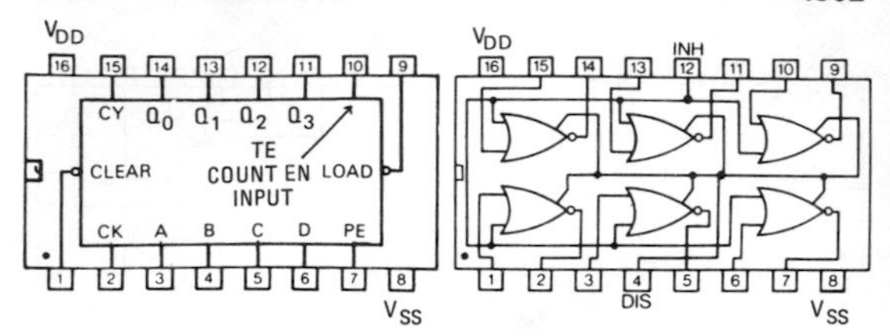

4508

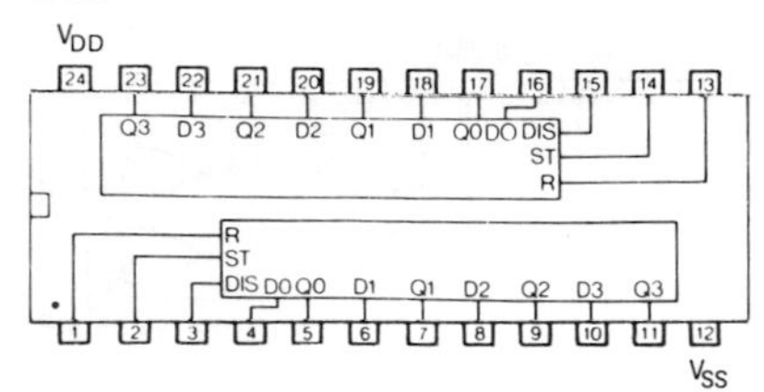

4510

4511

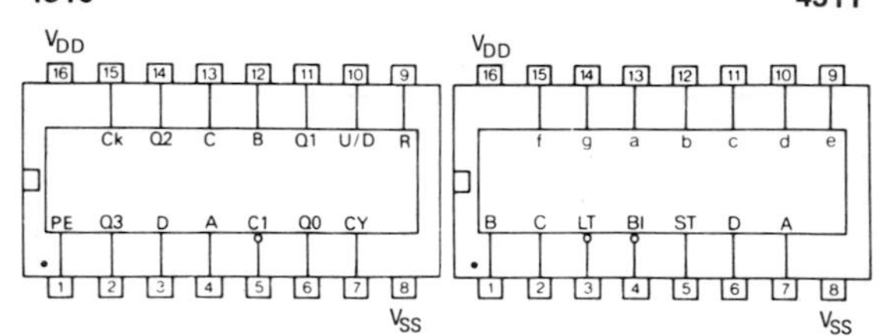

4512

4513

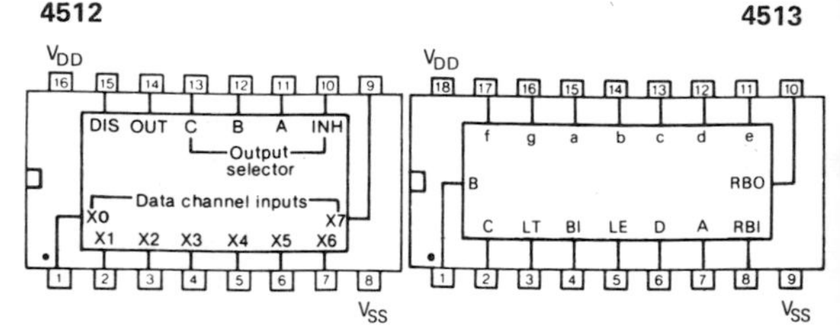

4514

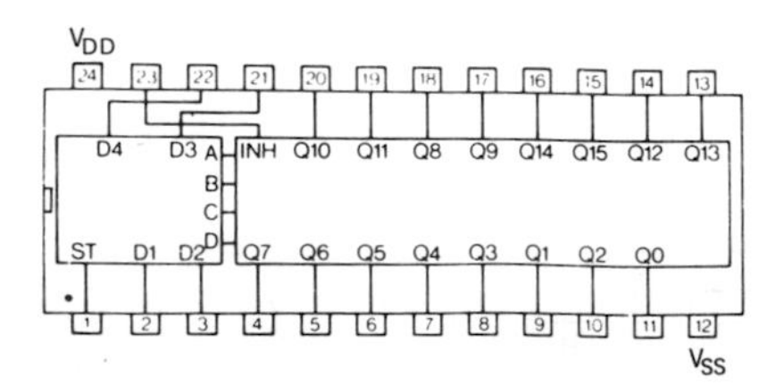

4515

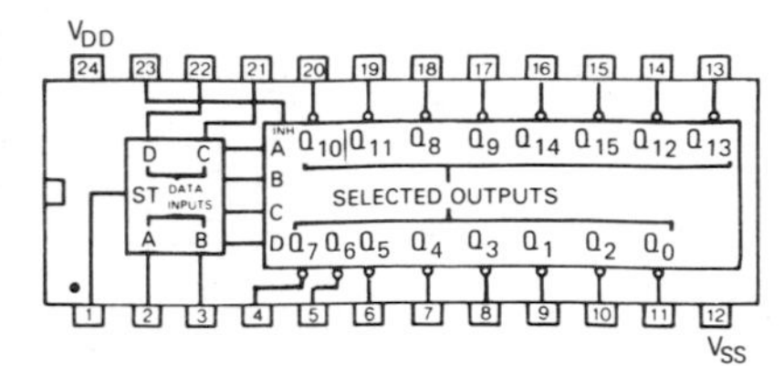

4516

4518

4519

4520

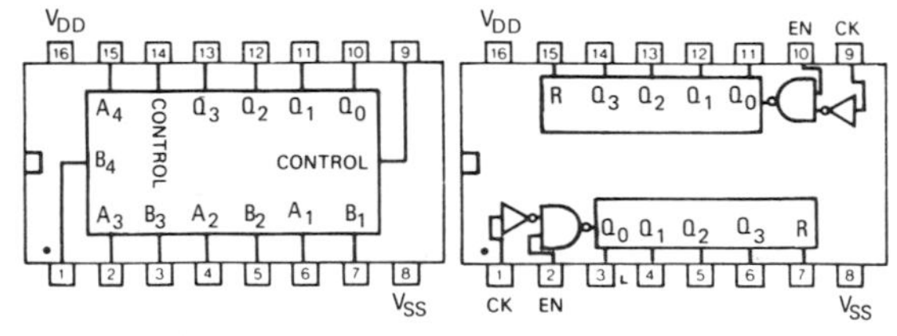

4521

4522

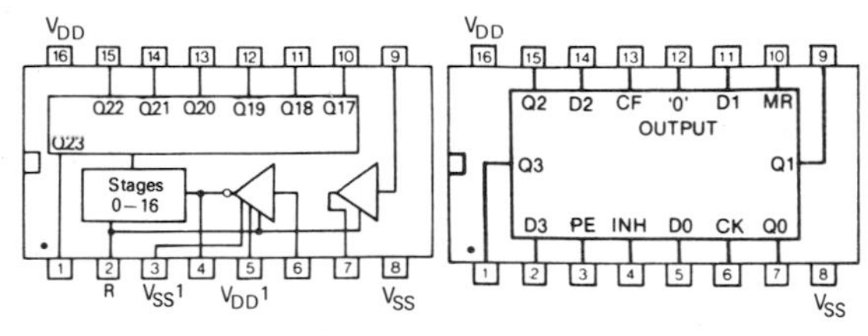

4526

4527

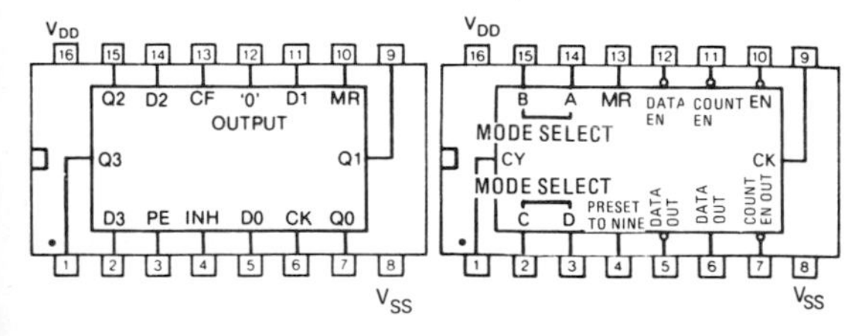

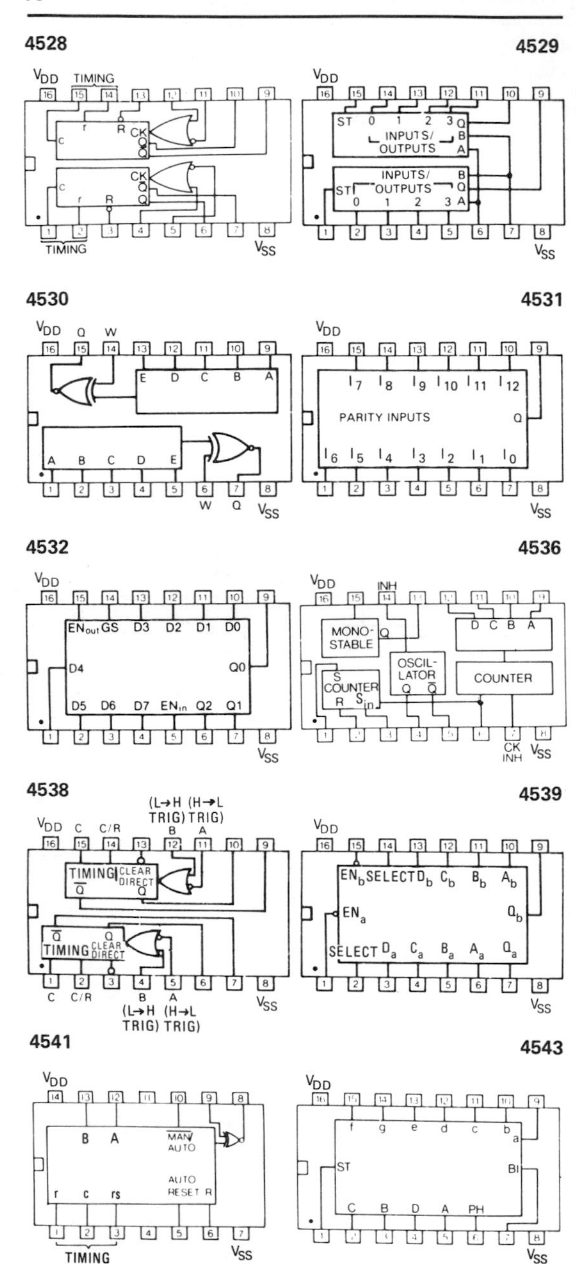
4528
VDD
TIMING
r
R
CK
Q
Q̄
c
CK
Q̄
Q
c
r
R
TIMING
VSS
4529
VDD
ST 0 1 2 3 Q
INPUTS/
OUTPUTS
B
A
INPUTS/
OUTPUTS
B
Q
ST
0 1 2 3 A
VSS
4530
VDD
Q
W
E D C B A
A B C D E
W
Q
VSS
4531
VDD
I7 I8 I9 I10 I11 I12
PARITY INPUTS
Q
I6 I5 I4 I3 I2 I1 I0
VSS
4532
VDD
ENout GS D3 D2 D1 D0
D4
Q0
D5 D6 D7 ENin Q2 Q1
VSS
4536
VDD
INH
MONO-
STABLE
Q
D C B A
OSCIL-
LATOR
Q Q̄
COUNTER
S
COUNTER
R Sin
CK
INH
VSS
4538
(L→H (H→L
TRIG) TRIG)
B A
VDD
C
C/R
TIMING
CLEAR
DIRECT
Q̄
Q
Q̄
Q
TIMING
CLEAR
DIRECT
C
C/R
B A
(L→H (H→L
TRIG) TRIG)
VSS
4539
VDD
ENb SELECT Db Cb Bb Ab
ENa
Qb
SELECT Da Ca Ba Aa Qa
VSS
4541
VDD
B A
MAN/
AUTO
AUTO
RESET R
r c rs
TIMING
VSS
4543
VDD
f g e d c b
a
ST
BI
C B D A PH
VSS

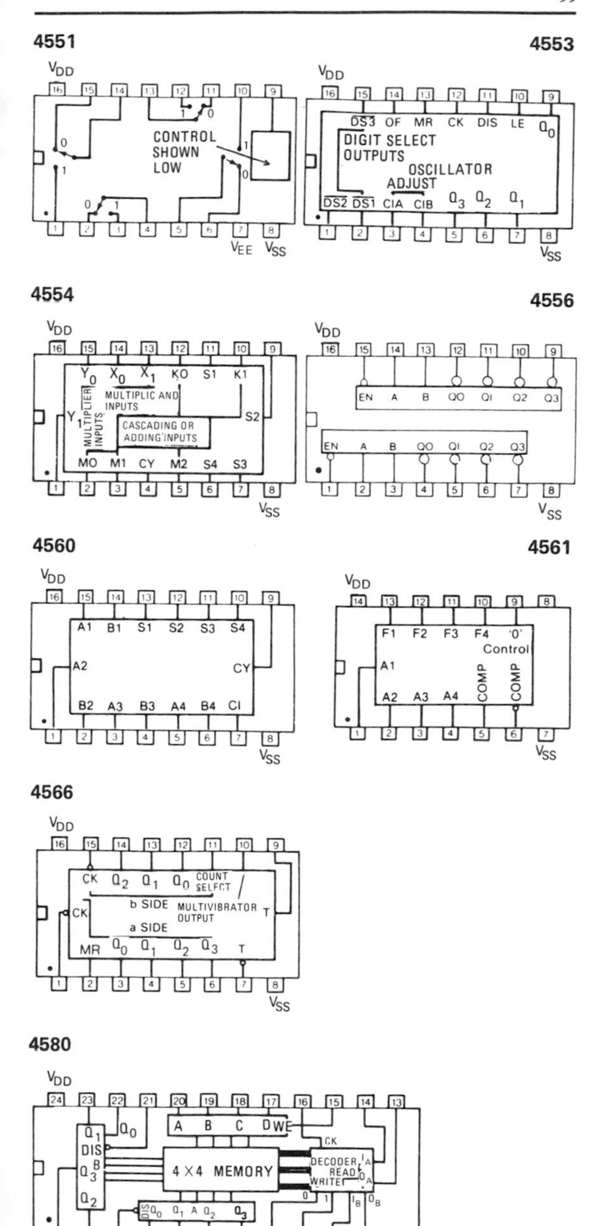
4551
4553
4554
4556
4560
4561
4566
4580
CONTROL SHOWN LOW
DIGIT SELECT OUTPUTS
OSCILLATOR ADJUST
MULTIPLIER INPUTS
MULTIPLIC AND INPUTS
CASCADING OR ADDING INPUTS
COUNT SELECT
b SIDE
a SIDE
MULTIVIBRATOR OUTPUT
4 X 4 MEMORY
DECODER
READ
WRITE

4581

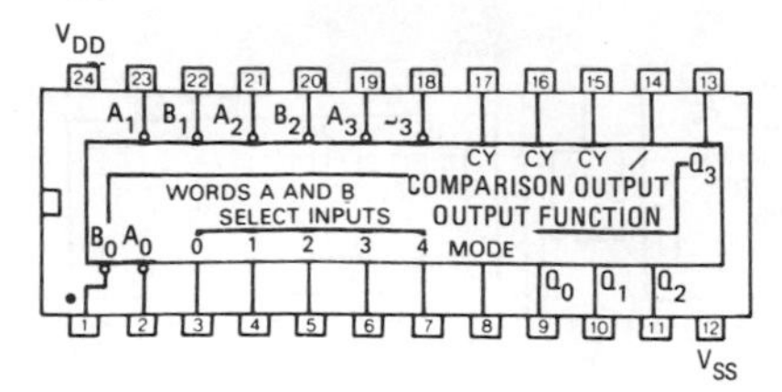

4582

4583

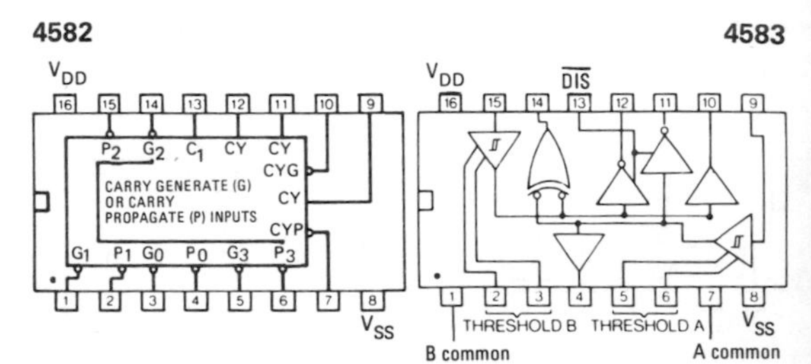

4585

VDD

Output

A3 B3 (A>B) (A<B) BO AO B1

B2 A2 (A=B) (A>B) (A<B) (A=B) A1

Output Input

VSS

4597

VDD

B C D E F G H

A

FULL FLAG

R X EN ST INC

VSS

4598

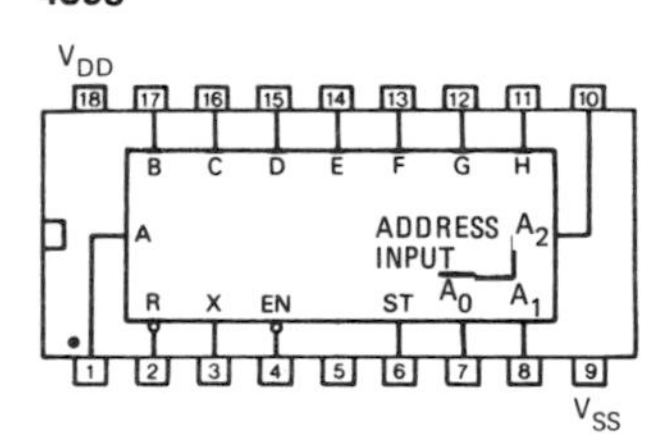

4599

45100

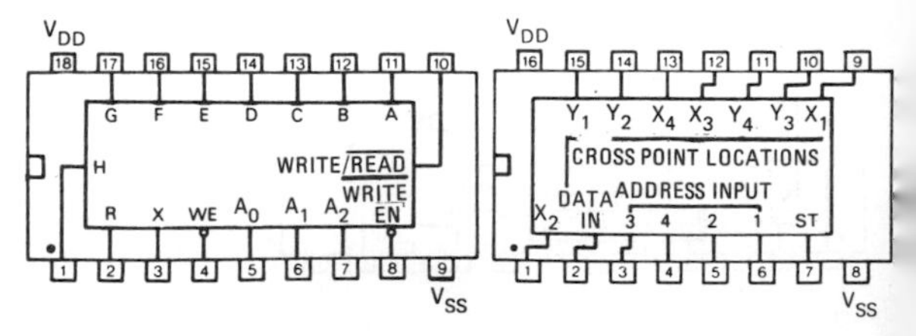

TTL and CMOS letter symbols

I_{IH} High level input current.
I_{IL} Low-level input current.
I_{OH} High-level output current.
I_O Off-state output current.
I_{OS} Short-circuit output current.
I_{CCH} Supply current output(s) high.
I_{CCL} Supply current output(s) low.
f_{max} Maximum clock frequency.
t_w Average pulse width.
I_{DD} Quiescent device current (CMOS).
I_{OL} Low level output current.
I_{IN} Input current.
I_{OZ} High impedance state output current of a 3-state output.
I_{CC} Quiescent device current (TTL).
t_h Hold time.
t_{PZX} Output enable time of a 3-state output to high or low level.
t_{PXZ} Output disable time of a 3-state output from high or low level.
t_{PD} Propagation delay time.
t_{TLH} Transition time from low to high level.
t_{THL} Transition time from high to low level.
Q_O Level of Q before the indicated steady-state input conditions were established.
$\overline{Q}_O$ Complement of Q_O.
V_{IH} High-level input voltage.
V_{IL} Low-level input voltage.
V_{T+} Positive-going threshold voltage.
V_{T-} Negative-going threshold voltage.
V_{OH} High-level output voltage.
V_{OL} Low-level output voltage.
$V_{O(ON)}$ On-state output voltage.
$V_{O(OFF)}$ Off-state output voltage.
V_{DD} DC supply voltage (CMOS).
V_{CC} DC supply voltage (TTL).
V_{SS} Ground (CMOS).
GND Ground (TTL).
V_{IN} Input voltage.
T_S Lead temperature when soldering.
P_D Package dissipation.
T_S Storage temperature range.
T_A Operating temperature range.
H High level (steady state).
L Low level (steady state).
↓ Transition from high to low.
↑ Transition from low to high level.
X Irrelevant input level.
Z High impedance state of a 3-state output.
⎍ One high level pulse.
⎍ (inverted) One low level pulse.
Toggle Each output changes to the complement of its previous level.
Q_n Level of Q before the most recent change.

ITU defined regions

For purposes of international allocations of frequencies the world has been divided into three regions as shown on the map.

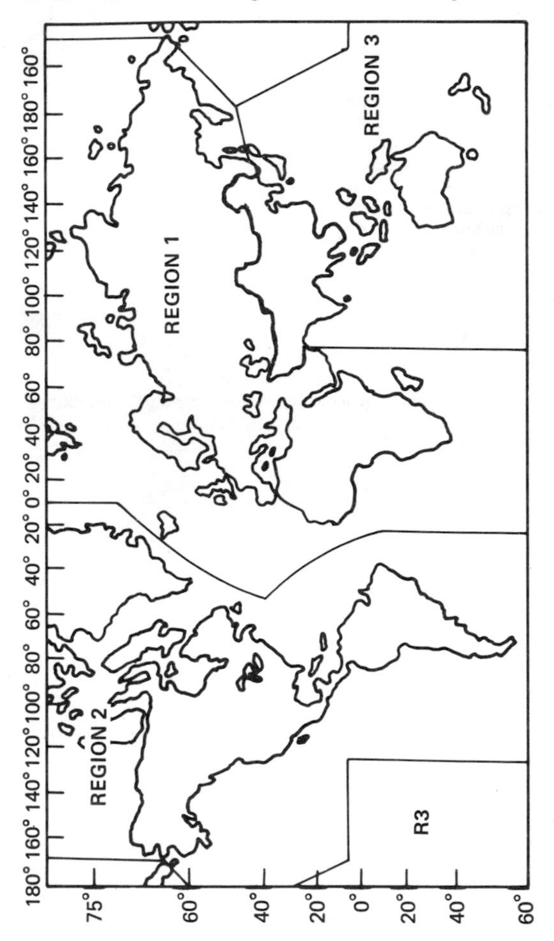

Designations of radio emissions

Radio emissions should be expressed in a three symbol code form, which defines the exact nature of carrier, signal and transmitted information. The first symbol defines the carrier, the second symbol defines the signal, and the third symbol defines the information.

First symbol

- A Double-sideband amplitude-modulated
- B Independent sideband amplitude-modulated
- C Vestigial sideband amplitude-modulated
- D Amplitude- and angle-modulated simultaneously, or in a pre-defined sequence
- F Frequency modulated
- G Phase modulated
- H Single-sideband, full carrier
- J Single-sideband, suppressed carrier
- K Amplitude-modulated pulse sequence
- L Width-modulated pulse sequence
- M Position phase modulated pulse sequence
- N Unmodulated carrier
- P Unmodulated pulse sequence
- Q Pulse sequence in which carrier is angle-modulated during the pulse period
- R Single-sideband, reduced or variable level carrier
- V Pulse sequence with a combination of carrier modulations, or produced by other means
- W Carrier is modulated by two or more of angle, amplitude, and pulse modes, simultaneously or in a defined sequence
- X Other cases

Second symbol

- 0 No modulating signal
- 1 Digital signal without modulating sub-carrier
- 2 Digital signal with modulating sub-carrier
- 3 Analogue signal
- 7 Two or more channels with digital signals
- 8 Two or more channels with analogue signals
- 9 Composite system with one or more channels of digital signals and one or more channels of analogue signals
- X Other cases

Third symbol

- A Aural telegraph
- B Automatic telegraph
- C Facsimile
- D Data
- E Telephony (and sound broadcasting)
- F Television
- N No information transmitted
- W Combination of any of the above
- X Other cases

Bandwidth and frequency designations

A four symbol code should be used to express bandwidth and frequency to three significant figures. A letter to denote the unit of frequency is placed in the position of the decimal point, where the letters and bandwidths are:

Letter	Bandwidth	Letter	Bandwidth
H	Below 1000 Hz	M	Between 1 and 999 MHz
K	Between 1 and 999 kHz	G	Between 1 and 999 GHz

So, a frequency of 120 Hz is 120H, while a frequency of 12 Hz is 12H0 etc.

Classes of radio stations

AL	Aeronautical radionavigation land station
AM	Aeronautical radionavigation mobile station
AT	Amateur station
AX	Aeronautical fixed station
BC	Broadcasting station, sound
BT	Broadcasting station, television
CA	Cargo ship
CO	Station open to official correspondence exclusively
CP	Station open to public correspondence
CR	Station open to limited public correspondence
CV	Station open exclusively to correspondence of a private agency
DR	Directive antenna provided with a reflector
EA	Space station in the amateur-satellite service
EB	Space station in the broadcasting-satellite service (sound broadcasting)
EC	Space station in the fixed-satellite service
ED	Space telecommand space station
EE	Space station in the standard frequency-satellite service
EF	Space station in the radiodetermination-satellite service
EG	Space station in the maritime mobile-satellite service
EH	Space research space station
EJ	Space station in the aeronautical mobile-satellite service
EK	Space tracking space station
EM	Meteorological-satellite space station
EN	Radionavigation-satellite space station
EO	Space station in the aeronautical radionavigation-satellite service
EQ	Space station in the maritime radionavigation-satellite service
ER	Space telemetering space station
ES	Station in the intersatellite service
EU	Space station in the land mobile-satellite service
EV	Space station in the broadcasting-satellite service (television)
EW	Space station in the earth exploration-satellite service
EX	Experimental station
EY	Space station in the time signal-satellite service
FA	Aeronautical station
FB	Base station
FC	Coast station
FL	Land station
FP	Port station
FR	Receiving station only, connected with the general network of telecommunication channels
FS	Land station established solely for the safety of life
FX	Fixed station
GS	Station on board a warship or a military or naval aircraft
LR	Radiolocation land station
MA	Aircraft station
ME	Space station
ML	Land mobile station
MO	Mobile station
MR	Radiolocation mobile station
MS	Ship station
ND	Non-directional antenna
NL	Maritime radionavigation land station
OD	Oceanographic data station

OE	Oceanographic data interrogating station
OT	Station open exclusively to operational traffic of the service concerned
PA	Passenger ship
RA	Radio astronomy station
RC	Non-directional radio beacon
RD	Directional radio beacon
RG	Radio direction-finding station
RM	Maritime radionavigation mobile station
RT	Revolving radio beacon
SM	Meteorological aids station
SS	Standard frequency and time signal station
TA	Space operation earth station in the amateur-satellite service
TB	Fixed earth station in the aeronautical mobile-satellite service
TC	Earth station in the fixed-satellite service
TD	Space telecommand earth station
TE	Transmitting earth station
TF	Fixed earth station in the radiodetermination-satellite service
TG	Mobile earth station in the maritime mobile-satellite service
TH	Earth station in the space research service
TI	Earth station in the maritime mobile-satellite service at a specified fixed point
TJ	Mobile earth station in the aeronautical mobile-satellite service
TK	Space tracking earth station
TL	Mobile earth station in the radiodetermination-satellite service
TM	Earth station in the meteorological-satellite service
TN	Earth station in the radionavigation-satellite service
TO	Mobile earth station in the aeronautical radionavigation-satellite service
TP	Receiving earth station
TQ	Mobile earth station in the maritime radionavigation-satellite service
TR	Space telemetering earth station
TS	Television, sound channel
TT	Earth station in the space operation service
TU	Mobile earth station in the land mobile-satellite service
TV	Television, vision channel
TW	Earth station in the earth exploration-satellite service
TX	Fixed earth station in the maritime radionavigation-satellite service
TY	Fixed earth station in the land mobile-satellite service
TZ	Fixed earth station in the aeronautical radionavigation-satellite service

Standard frequency transmissions

Frequency (kHz)	Wavelength (metres)	Station	Country	Power (kW)
2500	120	MSF Rugby (SF)	UK	0·5
		WWV(SF) Fort Collins	USA	2·5
		WWVH(SF) Kekaha	Hawaii	5

Frequency (kHz)	Wavelength (metres)	Station	Country	Power (kW)
2500	120	ZLF(SF) Wellington	New Zealand	—
		RCH(SF) Tashkent	USSR	1
3330	90.09	CHU(SF) Ottawa	Canada	3
4996	60.05	RWM(SF) Moscow	USSR	5
5000	60	WWV(SF) Fort Collins	USA	10
		WWVH(SF) Kekaha	Hawaii	10
		LOL(SF) Buenos Aires	Argentina	2
		MSF(SF) Rugby	UK	0·5
		IBF(SF) Turin	Italy	5
		RCH(SF) Tashkent	USSR	1
5004	59·95	RID(SF) Irkutsk	USSR	1
7335	40·90	CHU(SF) Ottawa	Canada	10
7500	40	VNG(SF) Lyndhurst	Australia	10
8167·5	36·73	LQB9(SF) Buenos Aires	Argentina	5
9996	30·01	RWM(SF) Moscow	USSR	5
10000	30	WWV(SF) Fort Collins	USA	10
		WWVH(S) Kekaha	Hawaii	10
		LOL(SF) Buenos Aires	Argentina	2
		MSF(SF) Rugby	UK	0·5
		RTA(SF) Novosibirsk	USSR	5
		RCH(SF) Tashkent	USSR	1
10004	29·99	RID(SF) Irkutsk	USSR	1
12000	25	VNG(SF) Lyndhurst	Australia	10
14670	20.45	CHU(SF) Ottawa	Canada	3
14996	20·01	RWM(SF) Moscow	USSR	8
15000	20	WWV(SF) Fort Collins	USA	10
		WWVH(SF) Kekaha	Hawaii	10
		LOL(SF) Buenos Aires	Argentina	2
		RTA(SF) Novosibirsk	USSR	5
15004	19·99	RID(SF) Irkutsk	USSR	1
16384	18·31	Allouis	France	2000
17550	17·09	LQC20(SF) Buenos Aires	Argentina	5
20000	15	WWV(SF) Fort Collins	USA	2·5

The electromagnetic wave spectrum

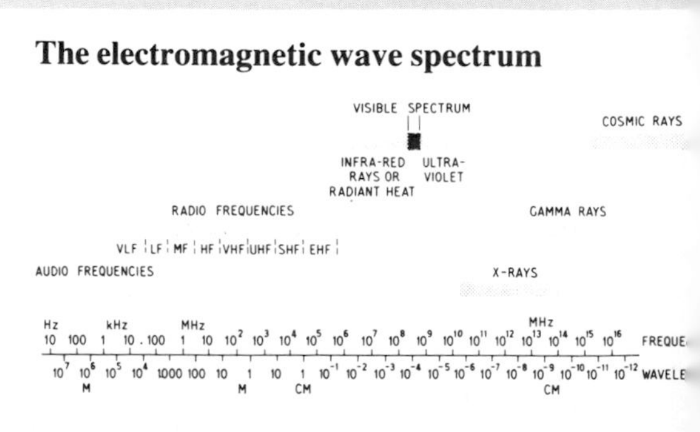

The ionosphere

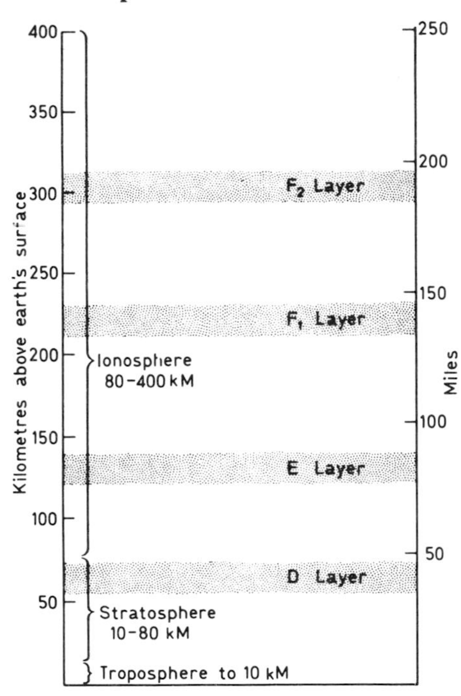

Radio wavebands

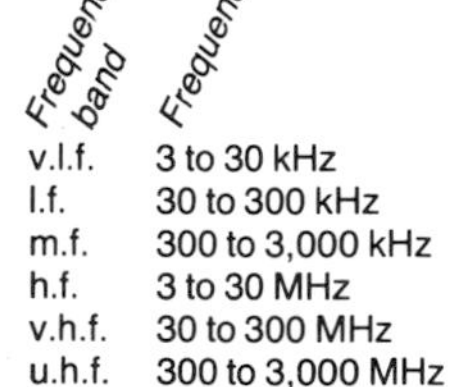

Frequency band	Frequency	Wavelength	Waveband definition
v.l.f.	3 to 30 kHz	100,000 to 10,000 m	myriametric
l.f.	30 to 300 kHz	10,000 to 1,000 m	kilometric
m.f.	300 to 3,000 kHz	1,000 to 100 m	hectometric
h.f.	3 to 30 MHz	100 to 10 m	decametric
v.h.f.	30 to 300 MHz	10 to 1 m	metric
u.h.f.	300 to 3,000 MHz	1 to 0·1 m	decimetric
s.h.f.	3 to 30 GHz	10 to 1 cm	centimetric
e.h.f.	30 to 300 GHz	1 to 0·1 cm	millimetric
e.h.f.	300 to 3,000 GHz	0·1 to 0·01 cm	decimillimetric

UK broadcasting bands

Frequency band	Frequency	Use
Long wave	150–285 kHz (2000–1053 m)	AM radio
Medium wave	525–1605 kHz (571–187 m)	AM radio
Band II (VHF)	88–97·6 MHz and 102–104·5 MHz	FM radio
Band IV (UHF)	470–582 MHz (channels 21 to 34)	TV
Band V (UHF)	614–854 MHz (channels 39 to 68)	TV
Band VI (SHF)	11·7–12·5 GHz (channels 1 to 40)	satellite TV

Band II is being extended to cover 88–108 MHz. However, due to existing use of the band at certain frequencies, this cannot be completed until 1990 (97·6–102 MHz) or 1996 (104·6–108 MHz).

The Beaufort scale

Force	Specification	Description	Speed (kmh^{-1})
0	Calm	Smoke rises vertically	Less than 1
1	Light air	Smoke drift shows wind direction	1–5
2	Light breeze	Wind can be felt on face	6–11
3	Gentle breeze	Leaves/twigs in constant motion	12–19
4	Moderate breeze	Dust blown about/small branches move	20–29
5	Fresh breeze	Small trees sway	30–39
6	Strong breeze	Large branches move	40–50
7	Near gale	Whole trees move, hard to walk	51–61
8	Gale	Twigs break, very hard to walk	62–74
9	Strong gale	Slight structural damage occurs, chimneys, slates blown off	75–87
10	Storm	Trees uprooted, considerable structural damage	88–101
11	Violent storm	Widespread damage	102–117
12	Hurricane	Catastrophic damage	>119

Boundaries of sea areas
As used in BBC and BT weather forecasts

Stations whose latest reports are broadcast in the 5 minute forecasts on Radio 2 (200 kHz) at 0033, 0633 and 1755 (daily), 1155 (Sundays) and 1355 (weekdays).

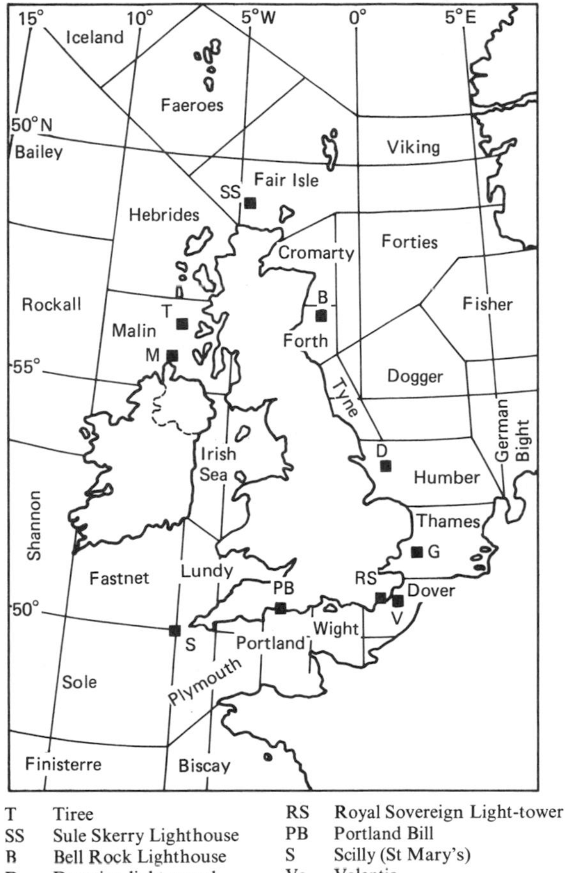

T	Tiree	RS	Royal Sovereign Light-tower
SS	Sule Skerry Lighthouse	PB	Portland Bill
B	Bell Rock Lighthouse	S	Scilly (St Mary's)
D	Dowsing light-vessel	Va	Valentia
G	Galloper light-vessel	R	Ronaldsway
V	Varne light-vessel	M	Malin Head Lighthouse

Overall rating for telephony

Symbol	*Operating condition*	*Quality*
5 Excellent	Signal quality unaffected	Commercial
4 Good	Signal quality slightly affected	
3 Fair	Signal quality seriously affected. Channel usable by operators or by experienced subscribers	Marginally commercial
2 Poor	Channel just usable by operators	Not commercial
1 Unusable	Channel unusable by operators	

The SINPFEMO code

	S	I	N	P
		Degrading effect of:		
Rating scale	Signal strength	Interference (QRM)	Noise (QRN)	Propagation disturbance
5	Excellent	Nil	Nil	Nil
4	Good	Slight	Slight	Slight
3	Fair	Moderate	Moderate	Moderate
2	Poor	Severe	Severe	Severe
1	Barely audible	Extreme	Extreme	Extreme

The SINPO code

	S	*I*	*N*	*P*	*O*
Rating scale	*Signal strength*	*Interference (QRM)*	*Noise (QRN)*	*Propagation disturbance*	*Overall readability (QRK)*
5	Excellent	Nil	Nil	Nil	Excellent
4	Good	Slight	Slight	Slight	Good
3	Fair	Moderate	Moderate	Moderate	Fair
2	Poor	Severe	Severe	Severe	Poor
1	Barely audible	Extreme	Extreme	Extreme	Unusable

The SIO code

	S	*I*	*O*
Rating scale	*Signal strength*	*Interference*	*Overall merit*
4	Good	Nil or very slight	Good
3	Fair	Moderate	Fair
2	Poor	Heavy	Unusable

The SIO code is based on the SINPO code but in a simplified form. Using the SIO code is perfectly acceptable, however.

F	*E*	*M*	*O*
	Modulation:		
Frequency of fading	Quality	Depth	Overall rating
Nil	Excellent	Maximum	Excellent
Slow	Good	Good	Good
Moderatc	Fair	Fair	Fair
Fast	Poor	Poor or nil	Poor
Very fast	Very poor	Continuously overmodulated	Unusable

BBC AM radio stations

Radio 1	*kHz*	*m*	*kW*
Barnstaple	1053	285	1
Barrow	1053	285	1
Bexhill	1053	285	2
Bournemouth	1485	202	2
Brighton	1053	285	2
Brookmans Park	1089	275	150
Burghead	1053	285	20
Droitwich	1053	285	150
Dundee	1053	285	1
Enniskillen	1053	285	1
Fareham	1089	275	1
Folkestone	1053	285	1
Hull	1053	285	1
Lisnagarvey	1089	275	10
Londonderry	1053	285	1
Moorside Edge	1089	275	150
Postwick	1053	285	10
Redmoss	1089	275	2
Redruth	1089	275	2
Stagshaw	1053	285	50
Start Point	1053	285	100
Tywyn	1089	275	1
Wallasey	1107	271	0·5
Washford	1089	275	50
Westerglen	1089	275	50
Whitehaven	1089	275	1

Radio 2	*kHz*	*m*	*kW*
Barrow	693	433	1
Bexhill	693	433	1
Bournemouth	909	330	1
Brighton	693	433	1
Brookmans Park	909	330	150
Burghead	693	433	50
Clevedon	909	330	50
Droitwich	693	433	150
Enniskillen	693	433	1
Exeter	909	330	1
Fareham	909	330	1
Folkestone	693	433	1
Lisnagarvey	909	330	10
Londonderry	909	330	1
Moorside Edge	909	330	200
Postwick	693	433	10
Redmoss	693	433	1
Redruth	909	330	2
Stagshaw	693	433	50
Start Point	693	433	50
Tywyn	990	303	1
Westerglen	909	330	50
Whitehaven	909	330	1

Radio 3	*kHz*	*m*	*kW*
Bournemouth	1197	251	0·5
Brighton	1215	247	1
Brookmans Park	1215	247	50
Burghead	1215	247	20
Cambridge	1197	251	0·2
Droitwich	1215	247	30
Enniskillen	1197	251	1
Fareham	1215	247	1
Hull	1215	247	0·3

	kHz	m	kW
Lisnagarvey	1215	247	10
Londonderry	1215	247	0·25
Moorside Edge	1215	247	100
Newcastle	1215	247	2
Plymouth	1215	247	1
Postwick	1215	247	1
Redmoss	1215	247	2
Redruth	1215	247	2
Torbay	1197	251	0·5
Tywyn	1215	247	0·5
Washford	1215	247	60
Westerglen	1215	247	50

Radio 4

	kHz	m	kW
Burghead	200	1500	50
Carlisle	1485	202	1
Droitwich	200	1500	400
Enniskillen	774	388	1
Lisnagarvey	720	417	10
London (Lots Road)	720	417	0·5
Londonderry	720	417	0·25
Newcastle	603	498	2
Plymouth	774	388	1
Redmoss	1449	207	2
Redruth	756	397	2
Westerglen	200	1500	50

Radio Scotland

	kHz	m	kW
Burghead	810	370	100
Dumfries	585	513	2
Redmoss	810	370	5
Westerglen	810	370	100

Radio Ulster

	kHz	m	kW
Enniskillen	873	344	1
Lisnagarvey	1341	224	100
Londonderry	792	379	1

Radio Wales

	kHz	m	kW
Forden	882	340	1
Llandrindod Wells	1125	267	1
Penmon	882	340	10
Tywyn	882	340	5
Washford	882	340	70
Wrexham	657	457	2

UK CB radio

27 MHz band: 27·60125 to 27·99125 MHz.

40 channels at 10 kHz spacing.

Max. e.r.p. 2W; max. transmitter output 4W.

Aerial: single rod or wire, 1·5m overall length, base loaded. If mounted higher than 7m, transmitter output to be reduced at least 10dB.

Modulation: F.M. only, deviation ±2·5 kHz max.

934 MHz band: 934·025 to 934·975 MHz.

20 channels at 50kHz (may be reduced to 25kHz later). If synthesizer used spacing may be 25kHz on precise channel frequencies specified.

Max. e.r.p. 25W; max. transmitter output 8W; if aerial integral, max. e.r.p. 3W.

Aerial: may have up to four elements, none exceeding 17cm. If mounted higher than 10m, transmitter output to be reduced at least 10dB.

Modulation: FM only, deviation ±5·0kHz max.

Spurious emissions: For both bands, not exceeding 0·25μW, except for specified frequency bands where the limit is 50nW.

For full specifications see publication MPT 1320 (27 MHz) and MPT 1321 (934 MHz) from HMSO.

BBC VHF/FM radio stations

Notes: **Stereo services:** all services are stereo except where (m) is shown against a frequency.
Polarisation: H indicates horizontal polarisation; M indicates mixed polarisation; V indicates vertical polarisation.

England, Isle of Man, and Channel Islands	Radio 1 & 2 (MHz)	Radio 3 (MHz)	Radio 4 (MHz)	Polarisation	Maximum effective radiated power (kW)
Belmont	88·8	90·9	93·1	M	25
Holme Moss	89·3	91·5	93·7	M	240
Kendal	88·7	90·9	93·1	M	0·04
Morecambe Bay	90·0	92·2	94·4	M	11
Oliver's Mount	89·9	92·1	94·3	H	0·25
Sheffield	89·9	92·1	94·3	M	0·32
Wensleydale	88·3	90·5	92·7	H	0·025
Wharfedale	88·4	90·6	92·8	M	0·04
Windermere	88·3	90·5	92·7	M	0·07
Les Platon (*Channel Isls.*)	91·1	94·75	97·1	M	16
North Hessary Tor	88·1	90·3	92·5	H	60
Okehampton	88·7	90·9	93·1	M	0·07
St. Thomas (Exeter)	89·0	91·2	93·4	M	0·055
Oxford	89·5	91·7	93·9	H	22
Peterborough	90·1	92·3	94·5	M	50
Cambridge	88·9	91·1	93·3	M	0·26
Pontop Pike	88·5	90·7	92·9	H	60
Chatton	90·1	92·3	94·5	M	5·6
Fenham		92·0	94·2	V	0·05
Weardale	89·7	91·9	94·1	H	0·1
Whitby	89·6	91·8	94·0	H	0·04
Redruth	89·7	91·9	94·1	H	9
Isles of Scilly	88·8	91·0	93·2	M	0·06
Rowridge	88·5	90·7	92·9	H	60
Ventnor	89·4	91·7	93·8	H	0·02
Sandale (*see also Scotland*)	88·1	90·3	94·7	H	120
Douglas (IOM)	88·4	90·6	92·8	M	12
Sutton Coldfield	88·3	90·5	92·7	M	240
Churchdown Hill	89·0	91·2	93·4	M	0·072
Hereford	89·7	91·9	94·1	H	0·026
Ludlow	89·6	91·8	94·0	M	0·01
Northampton	88·9	91·1	93·3	M	0·123
Swingate (Dover)	90·0	92·4	94·4	M	10
Tacolneston	89·7	91·9	94·1	M	240
Wenvoe (*see also Wales*)	89·9	92·1	94·3	M	240
Barnstaple	88·5	90·7	92·9	M	1
Bath	88·8	91·0	93·2	M	0·07
Ilchester Crescent	89·3	91.5	93·5	M	1·3
Wrotham	89·1	91·3	93·5	M	240
Brighton	90·1	92·3	94·5	M	0·5
Caterham	89·7	91·9	94·1	M	0·032
Guildford	88·1	90·3	92·5	M	3
Kenley	88·4	90·6	92·8	M	0·05
Winter Hill	88·6	90·8	93·0	M	4

Northern Ireland			Radio Ulster		
Divis	90·1	92·3	94·5	H	60
Ballycastle	88·8	91·0	93·2	M	0·05

	Radio 1 & 2 (MHz)	Radio 3 (MHz)	Radio Ulster (MHz)	Polarisation	Max. e.r.p. (kW)
Brougher Mountain	89·4	91·6	93·8	M	10
Kilkeel	89·4	91·6	93·8	H	0·025
Larne	89·1	91·3	93·5	M	0·1
Limavady	89·6	91·8	94·0	M	4
Maddybenny More	89·6	91·8	94·0	M	0·03
Rostrevor Forest	88·6(m)	90·8(m)	93·0(m)	M	0·064
Londonderry	88·7	90·9	93·1□	H	32

□ carries Radio Foyle(m)

Wales	Radio 1 & 2 (MHz)	Radio 3 (MHz)	Radio Cymru (MHz)	Radio Wales (MHz)		
Blaenplwyf	88·7	90·9	93·1		H	60
Dolgellau	90·1(m)	92·3(m)	94·5(m)		H	0·015
Ffestiniog	88·1	90·3	92·5		H	0·05
Llandyfriog	90·1	92·3	94·5		M	0·087
Machynlleth	89·4	91·6	93·8		H	0·06
Mynydd Pencarreg	89·7	91·9	94·1		M	0·4
Haverfordwest	89·3	91·5	93·7		H	10
Llanddona	89·8	92·0	94·2		M	50
Betws-y-Coed	88·2(m)	90·4(m)	92·6(m)		H	0·01
Llangollen	88·9	91·1	93·3(m)		M	22
Llandinam	90·1	92·3	94·5		H	0·02
Llanfyllin	89·1	91·3	93·5		M	0·014
Llanrhaeadr-ym-Mochant	89·8	92·0	94·2(m)		M	0·05
Long Mountain	89·6	91·8	94·0(m)		H	0·024
Wenvoe *(see also England)*	89·9	92·1	96·8		M	240
Aberdare	89·2	91·4	93·6		M	0·065
Abergavenny	88·6	90·8	93·0		H	0·017
Blaenavon	88·5	90·7	92·9	95·1(mg)	V	0·01
Brecon	88·9	91·1	93·3		H	0·01
Carmarthen	88·9	91·1	93·3		H	0·01
Carmel	88·4	90·6	92·8		M	3·2
Christchurch	—	—	—	103·0(mg)	M	0·5
Ebbw Vale	88·4	90·6	92·8		H	0·01
Kilvey Hill	89·5	91·7	93·9		M	1
Llandrindod Wells	89·1	91·3	93·5		H	1·5
Llanidloes	88·1	90·3	92·5		H	0·005
Pontypool	89·2	91·4	93·6		M	0·006
Varteg Hill	88·9	91·1	93·3		M	0·05

g carries Radio Gwent

Scotland			Radio Scotland (MHz)○	Radio Scotland (MHz)Ω		
Ashkirk	89·1	91·3	93·5●		H	18
Innerleithan	89·5	91·7	93·9●		M	0·02
Peebles	88·4	90·6	92.8●		M	0·02

	Radio 1 & 2 (MHz)	Radio 3 (MHz)	Radio Scotland (MHz) ○	Radio Scotland (MHz) Ω	Polarisation	Max. e.r.p. (kW)
Black Hill	89·9	92·1	94·3		M	240
Bressay	88·3	90·5	92·7⊕η‡		M	50
Darvel	89·5	91·7	93·9		M	10
Forfar	88·3	90·5	92·7		M	13·2
Pitlochry	89·2(m)	91·4(m)	93·6(m)		H	0·2
Fort William	89·3(m)	91·5(m)	93·7(m)†	98.9(m)	H	1·5
Ballachulish	88·1(m)	90·3(m)	92·5(m)†	97·7(m)	H	0·015
Glengorm	89·5(m)	91·7(m)	93·9(m)†	99·1(m)	M	2·2
Kinlochleven	89·7(m)	91·9(m)	94·1(m)†	99·3(m)	M	0·004
Mallaig	88·1(m)	90·3(m)	92·5(m)†	97·7(m)	H	0·02
Oban	88·9(m)	91·1(m)	93·3(m)†	98·5(m)	M	5
Keelylang Hill	89·3	91·5	93.7⊕η		M	40
Kirk o'Shotts	89·9	92·1	94·3		H	120
Ayr	88·7	90·9	93·1		H	0·055
Bowmore	88·1(m)	90·3(m)	92·5(m)		M	0·25
Campbeltown	88·4	90·6	92·8		M	0·4
Girvan	88·9	91·1	93·3		V	0·1
Lethanhill	88·3	90·5	92·7		M	0·2
Lochgilphead	88·3	90·5	92·7	97·9(m)	H	0·01
Millburn Muir	88·8	91·0	93·2		H	0·025
Perth	89·0(m)	91·2(m)	93·4(m)		H	0·012
Port Ellen	89·0(m)	91·2(m)	93·4(m)		M	0·2
Rosneath	89·2	91·4	93·6		H	0·025
Rothesay	88·5	90·7	92·9		M	0·6
South Knapdale	89·3	91·5	93·7	98·9(m)	H	1·1
Strachur	88·6	90·8	93·0	98·2(m)	M	0·02
Meldrum	88·7	90·9	93·1⊕		H	60
Durris	89·4	91·6	93·8⊕		M	2
Tullich	90·1	92·3	94·5⊕		M	0·042
Melvaig	89·1(m)	91·3(m)	93·5(m)†~	98·7(m)	H	22
Penifiler	89·5(m)	91·7(m)	93·9(m)†~	99·1(m)	H	0·006
Skriaig	88·5(m)	90·7(m)	92·9(m)†~	98·1(m)	H	10
Rosemarkie	89·6	91·8	94.0†		H	12
Grantown	89·8	92·0	94·6†		H	0·35
Kingussie	89·1	91·3	93·5†		H	0·035
Knock More	88·2	90·4	92·6†		M	0·5
Rumster Forest	90·1	92·3	94·5†		M	12·6
Sandale (*see also England*)	88·1	90·3	92·5*		H	120
Cambret Hill	88·7	90·9	93·1*		H	0·064
Stranraer	89·7	91·9	94·1		V	0·031

Ω Radio Scotland national service, also broadcast on 810 kHz medium frequency (reception of the medium frequency service is very poor in some parts of Western Scotland).

○ This service splits from the national Radio Scotland service at certain times to carry educational programmes and regional programmes.

⊕ carries Radio Aberdeen.

† carries Radio Highland.

‡ carries Radio Shetland.

η carries Radio Orkney.

~ carries Radio nan Eilean.

* carries Radio Solway.

● carries Radio Tweed.

BBC local radio stations

	Medium frequency (AM)		
	kHz	*metres*	*kW*
Bedfordshire	1161	258	0·08
Luton	630	476	0·3
Bristol	1548	194	5
Taunton	1323	227	1
Cambridgeshire	1026	292	0·5
Peterborough	1449	207	0·15
Cleveland	1548	194	1
Cornwall (Redruth)	630	476	2
Bodmin	657	457	0·5
Cumbria (Carlisle)	756	397	1
Whitehaven	1458	206	1
Derby	1116	269	0·5
Devon (Exeter)	990	303	1
Barnstaple	801	375	2
Plymouth	855	351	1
Torbay	1458	206	1
Essex	1530	196	0·1
Furness	837	358	1
Guernsey	1116	269	0·5
Humberside	1485	202	1
Jersey	1026	292	1
Kent (Hoo)	1035	290	1
Littlebourne	774	388	0·7
Rusthall	1602	187	0·25
Lancashire (Blackburn)	855	351	1
Oxcliffe	1557	193	0·25
Leeds	774	388	1
Leicester	837	358	0·7
Lincolnshire	1368	219	2
London	1458	206	50
Manchester	1458	206	5
Merseyside	1485	202	2
Newcastle	1458	206	2
Norfolk (Norwich)	855	351	1
West Lynn	873	344	0·25
Northampton	1107	271	0·5
Nottingham	1521	197	0·5
Clipstone	1584	189	1
Oxford	1485	202	0·5
Sheffield	1035	290	1
Shropshire	756	397	1
Woofferton	1584	189	0·3
Solent (Fareham)	999	300	1
Bournemouth	1359	221	0·25
Stoke-on-Trent	1503	200	0·5
Sussex (Brighton	1485	202	1
Bexhill	1161	258	1
Duxhurst	1368	219	0·5
WM (Birmingham)	1458	206	5
Wolverhampton	828	362	0·2
York (Fulford)	666	450	0·5
Scarborough	1260	238	0·5

	VHF (FM)		
	MHz	*kW*	*Polarisation*
Bedfordshire	103·8	20	Mixed
Sandy Heath	95·5	1·6	Mixed
Bristol (Main)	95·5	9	Mixed
Town	104·4	1	Mixed
Cambridgeshire			
Cambridge	96·0	1	Mixed
Peterborough	95·7	5	Mixed
Cleveland	96·6	5	Horizontal
Whitby	95·8	0·04	Horizontal
Cornwall (Redruth)	96·4	9	Horizontal
Caradon Hill	95·2	4·3	Mixed
Isles of Scilly	97·3	0·06	Mixed
Cumbria	95·6	5	Horizontal
Derby (Main)	104·5	5·5	Mixed
Town	94·2	0·01	Vertical
Devon: (N. Hessary Tor)	97·5	5	Horizontal
Exeter St. Thomas	97·0	0·4	Mixed
Huntshaw Cross	94·8	0·7	Mixed
Okehampton	96·2	0·07	Mixed
Essex	103·5	12·6	Mixed
Manningtree	104·9	5	Mixed
South Benfleet	95·3	2	Mixed
Furness	96·1	3·2	Mixed
Guernsey	93·2	1	Mixed
Humberside	95·9	9	Mixed
Jersey	88·8	4	Mixed
Kent (Wrotham)	96·7	9	Mixed
Dover	104·2	10	Mixed
Lancashire (Blackburn)	95·5	1·6	Mixed
Lancaster	103·3	2	Mixed
Winter Hill	103·9	2	Mixed
Leeds	92·4	5·2	Mixed
Wharfedale	95·3	0·04	Mixed
Leicester	95·1	0·3	Mixed
Lincolnshire	94·9	1·4	Mixed
London	94·9	2	Mixed
Manchester	95·1	4·2	Mixed
Merseyside	95·8	7·5	Mixed
Newcastle (Pontop Pike)	95·4	10	Mixed
Chatton	96·0	5·6	Mixed
Fenham	104·4	0·05	Mixed
Norfolk	95·1	5·7	Mixed
Great Massingham	96·7	4·2	Mixed
Northampton	104·2	4	Mixed
Geddington	103·6	0·8	Mixed
Nottingham	103·8	0·3	Mixed
Oxford	95·2	4·5	Horizontal
Sheffield (Main)	104·1	4·4	Mixed
Town	88·6	0·32	Mixed
Shropshire	96·0	5	Mixed
Ludlow	95·0	0·01	Mixed
Solent	96·1	5	Horizontal

	VHF(FM)		
	MHz	*kW*	*Polarisation*
Stoke-on-Trent	94·6	6	Mixed
Sussex (Brighton)	95·3	1	Mixed
Heathfield	104·5	10	Mixed
Reigate	104·0	4	Mixed
WM	95·6	11·4	Mixed
York	103·7	1·1	Vertical
Scarborough	97·2	0·25	Mixed

Independent local radio stations

	Medium wave			VHF	
	kHz	*m*	*kW*	*MHz*	*kW*
Aberdeen	1035	290	0·78	96·9	0·6
NorthSound					
Ayr (with Girvan)	1035	290	0·32	96·2	0·8
WestSound				97·1	0·15
Belfast	1026	293	1·7	97·4	1·0
Downtown Radio				102·4	
				96·4	
Birmingham	1152	261	3·0	94·8	2·0
BRMB Radio					
Bournemouth	828	362	0·27	97·2	1·0
2CR					
Bradford/Halifax	1278	235	0·43	97·5	0·5
& Huddersfield	1530	196	0·74	102·5	0·63
Pennine Radio					
Brighton	1323	227	0·5	103·4	0·9
Southern Sound					
Bristol	1260	238	1·6	96·3	0·6
GWR					
Bury St. Edmunds	1251	240	0·76	96·4	0·64
Saxon Radio					
Cardiff	1359	221	0·2	103·2	0·38
Red Dragon Radio					
Coventry	1359	220	0·27	97·0	0·62
Mercia Sound					
Doncaster	990	302	0·25	103·4	1·5
Radio Hallam					
Dundee/Perth	1161	258	0·7	95·8	0·5
Radio Tay	1584	189	0·21	96·4	0·25
East Kent	603	497	0·1	102·8	0·35
Invicta Radio				96·1	0·2
				97·0	0·5
				95·9	0·27
Edinburgh	1548	194	2·2	96·8	0·5
Radio Forth					
Exeter/Torbay	666	450	0·34	95·8	0·5
DevonAir Radio	954	314	0·32	95·1	0·25
Glasgow	1152	261	3·6	95·1	3·4
Radio Clyde					
Gloucester & Cheltenham	774	388	0·14	102·4	0·3

Severn Sound					
Great Yarmouth & Norwich	1152	261	0·83	97·6	3·3
Radio Broadland					
Guildford	1476	203	0·5	96·4	0·95
County Sound					
Hereford/Worcester	954	314	0·16	97·6	0·4
Radio Wyvern	1530	196	0·52	102·8	0·52
Humberside	1161	258	0·35	96·9	8·5
Viking Radio					
Inverness	1107	271	1·5	95·9	1·4
Moray Firth Radio					
Ipswich	1170	257	0·28	97·1	1·0
Radio Orwell					
Leeds	828	362	0·12	96·3	0·31
Radio Aire					
Leicester	1260	238	0·29	103·2	0·4
Leicester Sound					
Liverpool	1548	194	4·4	96·7	8·2
Radio City					
London - General & Entertainment	1548	194	97·5	95·8	2·0
Capital Radio					
London - News & Information	1152	261	23·5	97·3	2·0
LBC					
Chiltern Radio	792	379	0·27	96·9	0·5
Maidstone & Medway	1242	241	0·32	103·1	0·63
Invicta Radio					
Manchester	1152	261	1·5	103·0	4·0
Piccadilly Radio					
Newport (Gwent)	1305	230	0·2	97·4	0·5
Red Dragon Radio					
Northampton	1557	193	0·76	96·6	4·0
Hereward Radio					
Nottingham	999	301	0·25	96·2	0·3
Radio Trent					
Peterborough	1332	225	0·6	102·7	1·2
Hereward Radio					
Plymouth	1152	261	0·32	96·0	1·0
Plymouth Sound				96·6	
Portsmouth	1170	257	0·12	97·5	0·2
Ocean Sound	1557			103·2	
Preston & Blackpool	999	300	0·8	97·3	1·9
Red Rose Radio					
Reading	1431	210	0·14	97·0	0·5
Radio 210					
Reigate & Crawley	1521	197	0·64	102·7	3·6
Radio Mercury				97·5	0·006
Sheffield & Rotherham	1548	194	0·74	97·4	0·4
/Barnsley				96·1	0.05
Radio Hallam	1305	230	0·15	102·9	0·5
Southend/Chelmsford	1431	210	0·35	96·3	0·8
Essex Radio	1359	220	0·28	102·6	0·4
Stoke	1170	257	0·2	102·6	1·9
Signal Radio					
Swansea	1170	257	0·58	95·1	1·0

Swansea Sound					
Swindon/West Wilts.	1161	258	0·16	97·2	0·36
GWR	936	320	0·18	102·6	0·5
Teesside	1170	257	0·32	95·0	2·0
Radio Tees					
Tyne & Wear	1152	261	1·8	97·0	5·0
Metro Radio					
Wolverhampton & Black Country	990	303	0·09	97·2	1·0
Beacon Radio					
Wrexham & Deeside	1260	238	0·64	95·4	0·5
Marcher Sound/Sain-Y- Gororau					

BBC VHF test tone transmissions

Transmission starts about 4 minutes after the end of Radio 3 programmes on Mondays and Saturdays.

Time min.	Left channel	Right channel	Purpose
—	250 Hz at zero level	440 Hz at zero level	Identification of left and right channels and setting of reference level
2	900 Hz at +7 dB	900 Hz at +7 dB, antiphase to left channel	Adjustment of phase of regenerated subcarrier (see Note 4) and check of distortion with L-R signal only
6	900 Hz at +7 dB	900 Hz +7 dB, in phase with left channel	Check of distortion with L + R signal only
7	900 Hz at +7 dB	No modulation	Check of L to R cross-talk
8	No modulation	900 Hz at +7 dB	Check of R to L cross-talk
9	Tone sequence at −4 dB: 40 Hz 6·3 kHz, 100 Hz 10 kHz, 500 Hz 12·5 kHz, 1000 Hz 14 kHz. This sequence is repeated	No modulation	Check of L-channel frequency response and L to R cross-talk at high and low frequencies
11′40″	No modulation	Tone sequences as for left channel	Check of R-channel frequency response and R to L cross-talk at high and low frequencies
14′20″	No modulation	No modulation	Check of noise level in the presence of pilot

15′20″ End of test transmissions

Notes

1. This schedule is subject to variation or cancellation to accord with programme requirements and essential transmission tests.
2. The zero level reference corresponds to 40% of the maximum level of modulation applied to either stereophonic channel before pre-emphasis. All tests are transmitted with pre-emphasis.
3. Periods of tone lasting several minutes are interrupted momentarily at one-minute intervals.
4. With receivers having separate controls of subcarrier phase and crosstalk, the correct order of alignment is to adjust first the subcarrier phase to produce maximum output from either the L or the R channel and then to adjust the crosstalk (or 'separation') control for minimum crosstalk between channels.
5. With receivers in which the only control of crosstalk is by adjustment of subcarrier phase, this adjustment should be made on the crosstalk checks.
6. Adjustment of the balance control to produce equal loudness from the L and R loudspeakers is best carried out when listening to the announcements during a stereophonic transmission, which are made from a centre-stage position. If this adjustment is attempted during the tone transmissions, the results may be confused because of the occurrence of standing-wave patterns in the listening room.
7. The outputs of most receivers include significant levels of the 19-kHz tone and its harmonics, which may affect signal-level meters. It is important, therefore, to provide filters with adequate loss at these frequencies if instruments are to be used for the above tests.

Engineering information about broadcast services

Information about all BBC services as well as advice on how best to receive transmissions (including television) can be obtained from:

BBC Engineering Information Department
Broadcasting House
London
W1A 1AA

Telephone number (01) 927 5040

Transmitter service maps for most main transmitters can also be supplied, but requests for maps should be accompanied by a stamped addressed A4 sized envelope.

Similarly, information about all IBA broadcast services can be obtained from:

Engineering Information Service
Independent Broadcasting Authority
Crawley Court
Winchester
Hampshire
SO21 2QA

Telephone number (0962) 822444 or (01) 584 7011 and ask for engineering information

Relevant engineering information, including information regarding newly appointed transmitters etc., is broadcast by the IBA every Tuesday at 9.15 a.m. and 12.15 p.m., on Channel 4 television.

World time

Difference between local time and coordinated universal time
The differences marked + indicate the number of hours ahead of UTC. Differences marked − indicate the number of hours behind UTC. Variations from summer time during part of the year are decided annually and may vary from year to year.

	Normal time	Summer time
Afghanistan	+4½	+4½
Alaska	−9	−8
	−10	−9
Albania	+1	+2
Algeria	UTC	+1
Andorra	+1	+1
Angola	+1	+1
Anguilla	−4	−4
Antigua	−4	−4
Argentina	−3	−3
Ascension Isl.	UTC	UTC
Australia		
Victoria & New South Wales	+10	+11
Queensland	+10	+10
Tasmania	+10	+11
N. Territory	+9½	+9½
S. Australia	+9½	+10½
W. Australia	+8	+8
Austria	+1	+2
Azores	−1	UTC
Bahamas	−5	−4
Bahrain	+3	+3
Bangladesh	+6	+6
Barbados	−4	−4
Belau	+9	+9
Belgium	+1	+2
Belize	−6	−6
Benin	+1	+1
Bermuda	−4	−3
Bhutan	+6	+6
Bolivia	−4	−4
Botswana	+2	+2
Brazil		
(a) Oceanic Isl.	−2	−2
(b) Ea & Coastal	−3	−3
(c) Manaos	−4	−4
(d) Acre	−5	−5
Brunei	+8	+8
Bulgaria	+2	+3
Burkina Faso	UTC	UTC
Burma	+6½	+6½
Burundi	+2	+2
Cameroon	+1	+1
Canada		
(a) Newfoundland	−3½	−2½
(b) Atlantic (Labrador, Nova Scotia)	−4	−3
(c) Ea (Ontario, Quebec)	−5	−4
(d) Ce (Manitoba)	−6	−5
(e) Mountain (Alberta) NWT (Mountain)	−7	−6
(f) Pacific (Br. Columbia	−8	−7
Yukon	−8	−7
Canary Isl.	UTC	+1
Cape Verde Isl.	−1	−1
Cayman Isl.	−5	−4
Central African Republic	+1	+1
Chad	+1	+1
Chile	−4	−3
China People's Rep.	+8	+8
Christmas Isl.	+7	+7
Cocos Isl.	+6½	+6½
Colombia	−5	−5
Comoro Rep.	+3	+3
Congo	+1	+1
Cook Isl.	−10	−9½
Costa Rica	−6	−6
Cuba	−5	−4
Cyprus	+2	+3
Czechoslovakia	+1	+2
Denmark	+1	+2
Diego Garcia	+5	+5
Djibouti	+3	+3
Dominica	−4	−4
Dom. Rep.	−4	−4
Easter Isl.	−6	−5
Ecuador	−5	−5
Egypt	+2	+3
El Salvador	−6	−6
Equatorial Guinea	+1	+1
Ethiopia	+3	+3
Falkland Isl.	−4	−4
(Port Stanley)	−4	−3
Faroe Isl.	UTC	+1
Fiji	+12	+12
Finland	+2	+3
France	+1	+2
Gabon	+1	+1
Gambia	UTC	UTC
Germany	+1	+2
Ghana	UTC	UTC
Gibraltar	+1	+2

	Normal time	Summer time
Greece	+2	+3
Greenland		
Scoresbysund	−1	UTC
Thule area	−3	−3
Other areas	−3	−2
Grenada	−4	−4
Guadeloupe	−4	−4
Guam	+10	+10
Guatemala	−6	−6
Guiana (French)	−3	−3
Guinea (Rep.)	UTC	UTC
Guinea Bissau	UTC	UTC
Guyana (Rep.)	−3	−3
Haiti	−5	−4
Hawaii	−10	−10
Honduras (Rep.)	−6	−6
Hong Kong	+8	+8
Hungary	+1	+2
Iceland	UTC	UTC
India	+5½	+5½
Indonesia		
(a) Java, Bali, Sumatra	+7	+7
(b) Kalimantan, Sulawesi, Timor	+8	+8
(c) Moluccas, We. Irian	+9	+9
Iran	+3½	+3½
Iraq	+3	+4
Ireland	UTC	+1
Israel	+2	+3
Italy	+1	+2
Ivory Coast	UTC	UTC
Jamaica	−5	−4
Japan	+9	+9
Johnston Isl.	−10	−10
Jordan	+2	+3
Kampuchea	+7	+7
Kenya	+3	+3
Kiribati	+12	+12
Korea	+9	+9
Kuwait	+3	+3
Laos	+7	+7
Lebanon	+2	+3
Lesotho	+2	+2
Liberia	UTC	UTC
Libya	+1	+2
Lord Howe Isl.	+10½	+11½
Luxembourg	+1	+2
Macau	+8	+8
Madagascar	+3	+3
Madeira	UTC	+1
Malawi	+2	+2
Malaysia	+8	+8
Maldive Isl.	+5	+5
Mali	UTC	UTC
Malta	+1	+2
Marshall Isl.	+12	+12
Martinique	−4	−4
Mauritania	UTC	UTC
Mauritius	+4	+4

	Normal time	Summer time
Mayotte	+3	+3
Mexico		
(a) Campeche, Quintana Roo, Yucatan	−6	−5
(b) Sonora, Sinaloa, Nayarit, Baja' California Sur	−7	−7
(c) Baja California Norte	−8	−7
(d) other states	−6	−6
Micronesia		
Truk, Yap	+10	+10
Ponape	+11	+11
Midway Isl.	−11	−11
Monaco	+1	+2
Mongolia	+8	+9
Monserrat	−4	−4
Morocco	UTC	UTC
Mozambique	+2	+2
Nauru	+11½	+11½
Nepal	+5·45	+5·45
Netherlands	+1	+2
Neth. Antilles	−4	−4
New Caledonia	+11	+11
New Zealand	+12	+13
Nicaragua	−6	−6
Niger	+1	+1
Nigeria	+1	+1
Niue	−11	−11
Norfolk Isl.	+11½	+11½
N. Marianas	+10	+10
Norway	+1	+2
Oman	+4	+4
Pakistan	+5	+5
Panama	−5	−5
Papua N. Guinea	+10	+10
Paraguay	−4	−3
Peru	−5	−5
Philippines	+8	+8
Poland	+1	+2
Polynesia (Fr.)	−10	−10
Portugal	UTC	+1
Puerto Rico	−4	−4
Qatar	+3	+3
Reunion	+4	+4
Romania	+2	+3
Rwanda	+2	+2
Samoa Isl.	−11	−11
S. Tomé	UTC	UTC
Saudi Arabia	+3	+3
Senegal	UTC	UTC
Seychelles	+4	+4
Sierra Leone	UTC	UTC
Singapore	+8	+8
Solomon Isl.	+11	+11
Somalia	+3	+3
So. Africa	+2	+2
Spain	+1	+2
Sri Lanka	+5½	+5½
St. Helena	UTC	UTC

	Normal time	Summer time
St. Kitts-Nevis	−4	−4
St. Lucia	−4	−4
St. Pierre	−3	−3
St. Vincent	−4	−4
Sudan	+2	+2
Surinam	−3½	−3½
Swaziland	+2	+2
Sweden	+1	+2
Switzerland	+1	+2
Syria	+2	+3
Taiwan	+8	+8
Tanzania	+3	+3
Thailand	+7	+7
Togo	UTC	UTC
Tonga	+13	+13
Transkei	+2	+2
Trinidad & Tobago	−4	−4
Tristan da Cunha	UTC	UTC
Tunisia	+1	+1
Turks & Caicos	−4	−4
Turkey	+2	+3
Tuvalu	+12	+12
Uganda	+3	+3
United Arab Em.	+4	+4
United Kingdom	UTC	+1
Uruguay	−3	−3
USA		
(a) Eastern Zone*	−5	−4
(*) Indiana	−5	−5
(b) Central Zone	−6	−5
(c) Mountain Zone*	−7	−6
(*) Arizona	−7	−7
(d) Pacific Zone	−8	−7
USSR		
Moscow & Leningrad	+3	+4
Baku, Tbilisi	+4	+5
Sverdlovsk	+5	+6
Tashkent	+6	+7
Novobirsk	+7	+8
Irkutsk	+8	+9
Yakutsk	+9	+10
Khabarovsk	+10	+11
Magadan	+11	+12
Petropavlovsk	+12	+13
Anadyr	+13	+14
Vanuatu	+11	+12
Vatican	+1	+2
Venezuela	−4	−4
Vietnam	+7	+7
Virgin Isl.	−4	−4
Wake Isl.	+12	+12
Wallis & Futuna	+11	+11
Yemen	+3	+3
Yugoslavia	+1	+2
Zaire		
Kinshasa	+1	+1
Lubumbashi	+2	+2
Zambia	+2	+2
Zimbabwe	+2	+2

International allocation of call signs

The first character or the first two characters of a call sign indicate the nationality of the station using it.

AAA–ALZ	USA
AMA–AOZ	Spain
APA–ASZ	Pakistan
ATA–AWZ	India
AXA–AXZ	Australia
AYA–AZZ	Argentina
A2A–A2Z	Botswana
A3A–A3Z	Tonga
A5A–A5Z	Bhutan
BAA–BZZ	China
CAA–CEZ	Chile
CFA–CKZ	Canada
CLA–CMZ	Cuba
CNA–CNZ	Morocco
COA–COZ	Cuba
CPA–CPZ	Bolivia
CQA–CRZ	Portuguese Territories
CSZ–CUZ	Portugal
CVA–CXZ	Uruguay
CYA–CZZ	Canada
C2A–C2Z	Nauru
C3A–C3Z	Andorra
DAA–DTZ	Germany
DUA–DZZ	Philippines
EAA–EHZ	Spain
EIA–EJZ	Ireland
EKA–EKZ	USSR
ELA–ELZ	Liberia
EMA–EOZ	USSR
EPA–EQZ	Iran
ERA–ERZ	USSR

ESA–ESZ	Estonia (USSR)
ETA–ETZ	Ethiopia
EUA–EWZ	Belorussia (USSR)
EXA–EZZ	USSR
FAA–FZZ	France and Territories
GAA–GZZ	United Kingdom
HAA–HAZ	Hungary
HBA–HBZ	Switzerland
HCA–HDZ	Ecuador
HEA–HEZ	Switzerland
HFA–HFZ	Poland
HGA–HGZ	Hungary
HHA–HHZ	Haiti
HIA–HIZ	Dominican Republic
HJA–HKZ	Colombia
HLA–HMZ	Korea
HNA–HNZ	Iraq
HOA–HPZ	Panama
HQA–HRZ	Honduras
HSA–HSZ	Thailand
HTA–HTZ	Nicaragua
HUA–HUZ	El Salvador
HVA–HVZ	Vatican State
HWA–HYZ	France and Territories
HZA–HZZ	Saudi Arabia
IAA–IZZ	Italy and Territories
JAA–JSZ	Japan
JTA–JVZ	Mongolia
JWA–JXZ	Norway
JYA–JYZ	Jordan
JZA–JZZ	West Irian
KAA–KZZ	United States
LAA–LNZ	Norway
LOA–LWZ	Argentina
LXA–LXZ	Luxembourg
LYA–LYZ	Lithuania (USSR)
LZA–LZZ	Bulgaria
L2A–L9Z	Argentina
MAA–MZZ	United Kingdom
NAA–NZZ	United States
OAA–OCZ	Peru
ODA–ODZ	Lebanon
OEA–OEZ	Austria
OFA–OJZ	Finland
OKA–OMZ	Czechoslovakia
ONA–OTZ	Belgium
OUA–OZZ	Denmark
PAA–PIZ	Netherlands
PJA–PJZ	Netherlands West Indies
PKA–POZ	Indonesia
PPA–PYZ	Brazil
PZA–PZZ	Surinam
QAA–QZZ	(Service abbreviations)
RAA–RZZ	(USSR)
SAA–SMZ	Sweden
SNA–SRZ	Poland
SSA–SSM	Egypt
SSN–STZ	Sudan
SUA–SUZ	Egypt
SVA–SZZ	Greece
TAA–TCZ	Turkey
TDA–TDZ	Guatemala
TEA–TEZ	Costa Rica
TFA–TFZ	Iceland
TGA–TGZ	Guatemala
THA–THZ	France and Territories
TIA–TIZ	Costa Rica
TJA–TJZ	Cameroon
TKA–TKZ	France and Territories
TLA–TLZ	Central African Republic
TMA–TMZ	France and Territories
TNA–TNZ	Congo
TOA–TQZ	France and Territories
TRA–TRZ	Gabon
TSA–TSZ	Tunisia
TTA–TTZ	Chad
TUA–TUZ	Ivory Coast
TVA–TXZ	France and Territories
TYA–TYZ	Dahomey
TZA–TZZ	Mali
UAA–UQZ	USSR
URA–UTZ	Ukraine (USSR)
UUA–UZZ	USSR
VAA–VGZ	Canada
VHA–VNZ	Australia
VOA–VOZ	Canada
VPA–VSZ	British Territories
VTA–VWZ	India
VXA–VYZ	Canada
VZA–VZZ	Australia
WAA–WZZ	United States
XAA–XIZ	Mexico
XJA–XOZ	Canada
XPA–XPZ	Denmark
XQA–XRZ	Chile
XSA–XSZ	China
XTA–XTZ	Upper Volta
XUA–XUZ	Khmer Republic
XVA–XVZ	Vietnam
XWA–XWZ	Laos
XXA–XXZ	Portuguese Territories

XYA–XZZ	Burma
YAA–YAZ	Afghanistan
YBA–YHZ	Indonesia
YIA–YIZ	Iraq
YJA–YJZ	New Hebrides
YKA–YKZ	Syria
YLA–YLZ	Latvia (USSR)
YMA–YMZ	Turkey
YNA–YNZ	Nicaragua
YOA–YRZ	Romania
YSA–YSZ	El Salvador
YTA–YUZ	Yugoslavia
YVA–YYZ	Venezuela
YZA–YZZ	Yugoslavia
ZAA–ZAZ	Albania
ZBA–ZJZ	British Territories
ZKA–ZMZ	New Zealand
ZNA–ZOZ	British Territories
ZPA–ZPZ	Paraguay
ZQA–ZQZ	British Territories
ZRA–ZUZ	South Africa
ZVA–ZZZ	Brazil
2AA–2ZZ	United Kingdom
3AA–3AZ	Monaco
3BA–3BZ	Mauritius
3CA–3CZ	Equatorial Guinea
3DA–3DM	Swaziland
3DN–3DZ	Fiji
3EA–3FZ	Panama
3GA–3GZ	Chile
3HA–3UZ	China
3VA–3VZ	Tunisia
3WA–3WZ	Vietnam
3XA–3XZ	Guinea
3YA–3YZ	Norway
3ZA–3ZZ	Poland
4AA–4CZ	Mexico
4DA–4IZ	Phillipines
4JA–4LZ	USSR
4MA–4MZ	Venezuela
4NA–4OZ	Yugoslavia
4PA–4SZ	Sri Lanka
4TA–4TZ	Peru
4UA–4UZ	United Nations
4VA–4VZ	Haiti
4WA–4WZ	Yemen (YAR)
4XA–4XZ	Israel
4YA–4YZ	International Civil Aviation Organization
4ZA–4ZZ	Israel
5AA–5AZ	Libya
5BA–5BZ	Cyprus
5CA–5GZ	Morocco
5HA–5IZ	Tanzania
5JA–5KZ	Colombia
5LA–5MZ	Liberia
5NA–5OZ	Nigeria
5PA–5QZ	Denmark
5RA–5SZ	Malagasy Republic
5TA–5TZ	Mauretania
5UA–5UZ	Niger
5VA–5VZ	Togo
5WA–5WZ	Western Samoa
5XA–5XZ	Uganda
5YA–5ZZ	Kenya
6AA–6BZ	Egypt
6CA–6CZ	Syria
6DA–6JZ	Mexico
6KA–6NZ	Korea
6OA–6OZ	Somali Republic
6PA–6SZ	Pakistan
6TA–6UZ	Sudan
6VA–6WZ	Senegal
6XA–6XZ	Malagasy Republic
6YA–6YZ	Jamaica
6ZA–6ZZ	Liberia
7AA–7IZ	Indonesia
7JA–7NZ	Japan
7OA–7OZ	Yemen (PDRY)
7PA–7PZ	Lesotho
7QA–7QZ	Malawi
7RA–7RZ	Algeria
7SA–7SZ	Sweden
7TA–7YZ	Algeria
7ZA–7ZZ	Saudi Arabia
8AA–8IZ	Indonesia
8JA–8NZ	Japan
8OA–8OZ	Botswana
8PA–8PZ	Barbados
8QA–8QZ	Maldive Islands
8RA–8RZ	Guyana
8SA–8SZ	Sweden
8TA–8YZ	India
8ZA–8ZZ	Saudi Arabia
9AA–9AZ	San Marino
9BA–9DZ	Iran
9EA–9FZ	Ethiopia
9GA–9GZ	Ghana
9HA–9HZ	Malta
9IA–9JZ	Zambia
9KA–9KZ	Kuwait
9LA–9LZ	Sierra Leone
9MA–9MZ	Malaysia
9NA–9NZ	Nepal
9OA–9TZ	Zaire
9UA–9UZ	Burundi
9VA–9VZ	Singapore
9WA–9WZ	Malaysia
9XA–9XZ	Rwanda
9YA–9ZZ	Trinidad and Tobago

Amateur bands in the UK

The Schedule of frequency bands, powers, etc, which, for the sake of convenience, appear in an identical format in both the Class A and Class B licences

Frequency bands in MHz	*Status of allocations in the UK to: The Amateur Service*	*The Amateur Satellite Service*	*Maximum power Carrier*	*PEP*	*Permitted types of transmission*
1·810–1·850	Available to amateurs on a basis of non interference to other services		9dBW	15dBW	Morse Telephony RTTY Data Facsimile SSTV
1·850–2·000		No allocation			Morse Telephony Data Facsimile SSTV
3·500–3·800	Primary. Shared with other services	No allocation	20dBW	26dBW	Morse Telephony RTTY Data Facsimile SSTV
7·000–7·100	Primary	Primary			
10·100–10·150	Secondary	No allocation			
14·000–14·250	Primary	Primary			
14·250–14·350		No allocation			
18·068–16·168	Available to amateurs on a basis of non interference to other services. Antennas limited to horizontal polarisation, maximum gain 0dB with respect to a half-wave dipole	No allocation	10dBW	—	Morse, AIA only
21·000–21·450	Primary	Primary	20dBW	26dBW	Morse Telephony RTTY Data Facsimile SSTV
24·890–24·990	Available to amateurs on basis of non interference to other services. Antennas limited to horizontal polarisation, maximum gain 0dB with respect to a half-wave dipole	No allocation	10dBW	—	Morse, AIA only
28·000–29·700	Primary	Primary	20dBW	26dBW	Morse Telephony
50·000–50·500	**Primary**	**No allocation**	**14dBW**	**20dBW**	**Morse** **Telephony**
70·025–70·500	Secondary basis until further notice. Subject to not causing interference to other services. Use of any frequency shall cease immediately on demand of a government official	No allocation	16dBW	22dBW	RTTY Data Facsimile SSTV
144·0–146·0*	Primary	Primary	20dBW	26dBW	
430·0–431.0	Secondary. This band is not available for use within the area bounded by: 53 N 02 E, 55 N 02 E, 53 N 03 W, and 55 N 03 W	No allocation	10dBW e.r.p.	16dBW e.r.p.	Morse Telephony RTTY Data Facsimile SSTV Television

431·0–432·0	Secondary. This band is not available for use: a) Within the area bounded by: 53 N 02 E, 55 N 02 E, 55 N 03 W, and 55 N 03 W. b) Within a 100km radius of Charing Cross. 51 30'30''N 00.07'24''W	No allocation	10dBW e.r.p.	16dBW e.r.p.	Morse Telephony RTTY Data Facsimile SSTV Television
432·0–435·0	Secondary	No allocation	20dBW	26dBW	
435·0–438·0		Secondary			
438·0–440·0		No allocation			
1240–1260	Secondary	No allocation			
1260–1270		Secondary Earth to Space only			
1270–1325		No allocation			
2310–2400					
2400–2450	Secondary. Users must accept interference from the ISM allocations in this band	Secondary. Users must accept interference from the ISM allocations in this band			
3400–3475	Secondary	No allocation			
5650–5670		Secondary Earth to Space only			
5670–5680					
5755–5765	Secondary. Users must accept interference from the ISM allocations in this band	No allocation			
5820–5830					
5830–5850		Secondary. Users must accept interference from the ISM allocations in this band. Space to Earth only			
10000–10450	Secondary	No allocation			
10450–10500		Secondary			
24000–24050	Primary. Users must accept interference from the ISM allocations in this band	Primary. Users must accept interference from the ISM allocations in this band			
24050–24250	Secondary. This band may only be used with the written consent of the Secretary of State. Users must accept interference from the ISM allocations in this band	No allocation			
47000–47200	Primary	Primary			
75500–76000					
142000–144000					
248000–250000					

*Except in accordance with clause 1(2)(c)(ii) holders of the Amateur Radio Licence (B) are not permitted to use frequencies below 144 MHz, nor may they use the type of transmission known as morse (whether sent manually or automatically).

Dipole lengths for the amateur bands

Amateur band (metres)	*Dipole length (metres)*
80	39
40	20·2
20	10·1
15	6·7
10	5·0

Amateur radio emission designations

The first symbol specifies the modulation of the main carrier, the second symbol the nature of the signal(s) modulating the main carrier, and the third symbol the type of information to be transmitted.

Amplitude modulation

A1A Telegraphy by on-off keying without the use of a modulating audio frequency

A1B Automatic telegraphy by on-off keying, without the use of a modulating audio frequency

A2A Telegraphy by on-off keying of an amplitude modulating audio frequency or frequencies, or by on-off keying of the modulated emission

A2B Automatic telegraphy by on-off keying of an amplitude modulating audio frequency or modulated emission

A3E Telephony, double sideband

A3C Facsimile transmission

H3E Telephony using single sideband full carrier, amplitude modulation

R3E Telephony, single sideband, reduced carrier

J3E Telephony, single sideband, suppressed carrier

A3F/
C3F Slow scan and high definition television

Frequency modulation

F1A Telegraphy by frequency shift keying without the use of a modulating frequency: one of two frequencies being emitted at any instant

F1B Automatic telegraphy by frequency shift keying without the use of a modulating frequency

F2A Telegraphy by on-off keying of a frequency modulating audio frequency or on-off keying of an f.m. emission

F2B Automatic telegraphy by on-off keying of a frequency modulating audio frequency or of an f.m. emission

F3E Telephony

F3C Facsimile transmission

F3F Slow scan and high definition television

Microwave band designation systems

MOD. discontinued system	GHz	*UK IEE recommended system*	GHz	*New NATO designation system*	GHz	*USA system*	GHz
P	0·08–0·39	A	0–0·25	L	1–2	P	0·225–0·39
L_2	0·39–1·0	B	0·25–0·5	S	2–4	L	0·39–1·55
L_1	1·0–2·5	C	0·5–1·0	C	4–8	S	1·55–5·2
S	2·5–4·1	D	1·0–2	X	7–12	X	5·2–10·9
C	4·1–7·0	E	2–3	J	12–18	K	10·9–36
X	7·0–11·5	F	3–4	K	18–26	Q	36–46
J	11·5–18·0	G	4–6	Q	26–40	V	46–56
K	18–33	H	6–8	V	40–60	W	56–100
Q	33–40	I	8–10	O	60–90		
O	40–60	J	10–20				
V	60–90	K	20–40				
		L	40–60				
		M	60–100				

International 'Q' code

Abbrev.	Question	Answer for advice
QRA	What is the name of your station?	The name of my station is ...
QRB	How far approximately are you from my station?	The approximate distance is ... miles
QRD	Where are you bound and where are you from?	I am bound for ... from ...
QRG	Will you tell me my exact frequency in kHz?	Your exact frequency is ... kHz.
QRH	Does my frequency vary?	Your frequency varies.
QRI	Is my note good?	Your note varies.
QRJ	Do you receive me badly? Are my signals weak?	I cannot receive you. Your signals are too weak.
QRK	Do you receive me well? Are my signals good?	I receive you well. Your signals are good.
QRL	Are you busy?	I am busy. Please do not interfere.
QRM	Are you being interfered with?	I am being interfered with.
QRN	Are you troubled by atmospherics?	I am troubled by atmospherics.
QRO	Shall I increase power?	Increase power.
QRP	Shall I decrease power?	Decrease power.
QRQ	Shall I send faster?	Send faster (... words per minute).
QRS	Shall I send more slowly?	Send more slowly (... words per minute).
QRT	Shall I stop sending?	Stop sending.
QRU	Have you anything for me?	I have nothing for you.
QRV	Are you ready?	I am ready.

QRX	Shall I wait? When will you call me again?	Wait (or wait until I have finished communicating with ...). I will call you at ... GMT.
QRZ	Who is calling me?	You are being called by ...
QSA	What is the strength of my signals? (1 to 5)	The strength of your signals is ... (1 to 5).
QSB	Does the strength of my signals vary?	The strength of your signals varies.
QSD	Is my keying correct? Are my signals distinct?	Your keying is indistinct. Your signals are bad.
QSL	Can you give me acknowledgement of receipt?	I give you acknowledgement of receipt.
QSM	Shall I repeat the last telegram (message) I sent you?	Repeat the last telegram (message) you have sent me.
QSO	Can you communicate with ... direct (or through the medium of ...)?	I can communicate with ... direct (or through the medium of ...).
QSP	Will you relay to ...?	I will relay to ...
QSV	Shall I send a series of V's?	Send a series of V's.
QSX	Will you listen for ... (call sign) on ... kHz?	I am listening for ... (call sign) on ... kHz.
QSZ	Shall I send each word or group twice?	Send each word or group twice.
QTH	What is your position in latitude and longitude?	My position is ... latitude ... longitude.
QTR	What is the exact time?	The exact time is ...

QSA Code (signal strength)

QSA1 . Hardly perceptible; unreadable.
QSA2 . Weak, readable now and then.
QSA3 . Fairly good; readable, but with difficulty.
QSA4 . Good; readable.
QSA5 . Very good; perfectly readable.

QRK Code (audibility)

R1 . Unreadable.
R2 . Weak signals; barely readable.
R3 . Weak signals; but can be copied
R4 . Fair signals; easily readable.
R5 . Moderately strong signals.
R6 . Good signals.
R7 . Good strong signals.
R8 . Very strong signals.
R9 . Extremely strong signals.

RST Code (readability)

1 . Unreadable.
2 . Barely readable, occasional words distinguishable.
3 . Readable with considerable difficulty.
4 . Readable with practically no difficulty.
5 . Perfectly readable.

(Signal strength)

1 . Faint, signals barely perceptible.
2 . Very weak signals.
3 . Weak signals.
4 . Fair signals.
5 . Fairly good signals.
6 . Good signals.
7 . Moderately strong signals.
8 . Strong signals.
9 . Extremely strong signals.

(Tone)

1 . Extremely rough hissing note.
2 . Very rough AC note, no trace of musicality.
3 . Rough, low-pitched AC note, slightly musical.
4 . Rather rough AC note, moderately musical.
5 . Musically modulated note.
6 . Modulated note, slight trace of whistle.
7 . Near DC note, smooth ripple.
8 . Good DC note, just a trace of ripple.
9 . Purest DC note.

(If the note appears to be crystal-controlled add an X after the appropriate number.)

International Morse Code

A	dit dah	·–	N	dah dit	–·
B	dah dit dit dit	–···	O	dah dah dah	–––
C	dah dit dah dit	–·–·	P	dit dah dah dit	·––·
D	dah dit dit	–··	Q	dah dah dit dah	––·–
E	dit	·	R	dit dah dit	·–·
F	dit dit dah dit	··–·	S	dit dit dit	···
G	dah dah dit	––·	T	dah	–
H	dit dit dit dit	····	U	dit dit dah	··–
I	dit dit	··	V	dit dit dit dah	···–
J	dit dah dah dah	·–––	W	dit dah dah	·––
K	dah dit dah	–·–	X	dah dit dit dah	–··–
L	dit dah dit dit	·–··	Y	dah dit dah dah	–·––
M	dah dah	––	Z	dah dah dit dit	––··

Number code

1	dit dah dah dah dah	·––––	6	dah dit dit dit dit	–····
2	dit dit dah dah dah	··–––	7	dah dah dit dit dit	––···
3	dit dit dit dah dah	···––	8	dah dah dah dit dit	–––··
4	dit dit dit dit dah	····–	9	dah dah dah dah dit	––––·
5	dit dit dit dit dit	·····	0	dah dah dah dah dah	–––––

Note of interrogation	dit dit dah dah dit dit	··––··
Note of exclamation	dah dah dit dit dah dah	––··––
Apostrophe	dit dah dah dah dah dit	·––––·
Hyphen	dah dit dit dit dit dah	–····–
Fractional bar	dah dit dit dah dit	–··–·
Brackets	dah dit dah dah dit dah	–·––·–
Inverted commas	dit dah dit dit dah dit	·–··–·
Underline	dit dit dah dah dit dah	··––·–
Prelim. call	dah dit dah dit dah	–·–·–
Break sign	dah dit dit dit dah	–···–
End message	dit dah dit dah dit	·–·–·
Error	dit dit dit dit dit dit	······

Timing

The basic timing measurement is the dot pulse (dit), all other morse code timings are a function of this unit length:

Dot length (dit)	one unit
Dash length (dah)	three units
Pause between elements of one character	one unit
Pause between characters	three units
Pause between words	seven units

Phonetic alphabet

To avoid the possibility of the letters of a call-sign being misunderstood, it is usual to use the words given below in place of the letters. For example, G6PY would be given as G6 Papa Yankee.

Letter	Code word	Pronunciation
A	Alfa	*AL* FAH
B	Bravo	*BRAH* VOH
C	Charlie	*CHAR* LEE
D	Delta	*DELL* TAH
E	Echo	*ECK* OH
F	Foxtrot	*FOKS* TROT
G	Golf	GOLF
H	Hotel	HOH *TELL*
I	India	*IN* DEE AH
J	Juliett	*JEW* LEE *ETT*
K	Kilo	*KEY* LOH
L	Lima	*LEE* MAH
M	Mike	MIKE
N	November	NO *VEM* BER
O	Oscar	*OSS* CAH
P	Papa	PAH *PAH*
Q	Quebec	KEH *BECK*
R	Romeo	*ROW* ME OH
S	Sierra	SEE *AIR* RAH
T	Tango	*TANG* GO
U	Uniform	*YOU* NEE FORM
V	Victor	*VIK* TAH
W	Whiskey	*WISS* KEY
X	X-ray	*ECKS* RAY
Y	Yankee	*YANG* KEY
Z	Zulu	*ZOO* LOO

Syllables in italic carry the accent.

Miscellaneous international abbreviations

C	Yes
N	No
W	Word
AA	All after ...
AB	All before ...
AL	All that has just been sent
BN	All between
CL	I am closing my station
GA	Resume sending
MN	Minute/minutes
NW	I resume transmission
OK	Agreed
UA	Are we agreed?
WA	Word after ...
WB	Word before ...
XS	Atmospherics

Amateur abbreviations

ABT	About
AGN	Again
ANI	Any
BA	Buffer amplifier
BCL	Broadcast listener
BD	Bad
BI	By
BK	Break in
BN	Been
CK	Check
CKT	Circuit
CLD	Called
CO	Crystal oscillator
CUD	Could
CUL	See you later
DX	Long distance
ECO	Electron-coupled oscillator
ES	And

FB	Fine business (good work)
FD	Frequency doubler
FM	From
GA	Go ahead, or Good afternoon
GB	Good-bye
GE	Good evening
GM	Good morning
GN	Good night
HAM	Radio amateur
HI	Laughter
HR	Hear or here
HRD	Heard
HV	Have
LTR	Later
MILS	Milliamperes
MO	Meter oscillator
ND	Nothing doing
NIL	Nothing
NM	No more
NR	Number
NW	Now
OB	Old boy
OM	Old man
OT	Old timer
PA	Power amplifier
PSE	Please
R	Received all sent
RAC	Rectified AC
RCD	Received
RX	Receiver
SA	Say
SED	Said
SIGS	Signals
SIGN	Signature
SSS	Single signal superheterodyne
SKD	Schedule
TKS	Thanks
TMN	Tomorrow
TNX	Thanks
TPTG	Tuned plate tuned grid
TX	Transmitter
U	You
UR	You are
VY	Very
WDS	Words
WKG	Working
WL	Will
WUD	Would
WX	Weather
YF	Wife
YL	Young lady
YR	Your
73	Kind regards
88	Love and kisses

Characteristics of world television systems

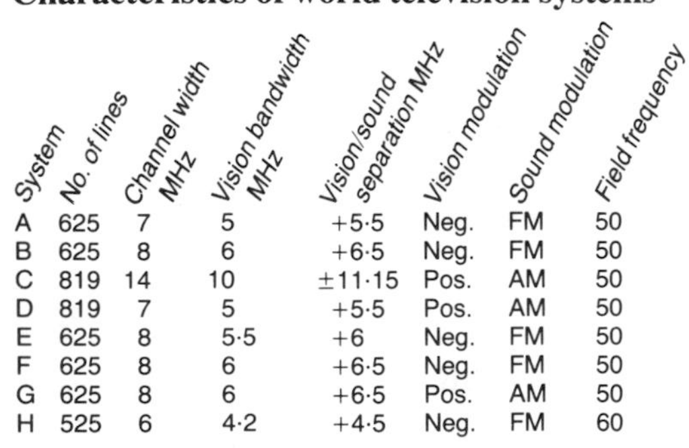

System	No. of lines	Channel width MHz	Vision bandwidth MHz	Vision/sound separation MHz	Vision modulation	Sound modulation	Field frequency
A	625	7	5	+5·5	Neg.	FM	50
B	625	8	6	+6·5	Neg.	FM	50
C	819	14	10	±11·15	Pos.	AM	50
D	819	7	5	+5·5	Pos.	AM	50
E	625	8	5·5	+6	Neg.	FM	50
F	625	8	6	+6·5	Neg.	FM	50
G	625	8	6	+6·5	Pos.	AM	50
H	525	6	4·2	+4·5	Neg.	FM	60

A—Most of Western Europe, Australia, New Zealand
B—USSR and Eastern Europe except East Germany
C—France, Monaco
D—Luxembourg
E—UK and Eire
F—French Overseas Territories
G—France
H—USA, most of Central and South America, Japan and others

UK 625-line television channels, bands IV and V

Channel no.	Frequencies (MHz)	Channel no.	Frequencies (MHz)	Channel no.	Frequencies (MHz)
21	470–478	37	598–606	53	726–734
22	478–486	38	606–614	54	734–742
23	486–494	39	614–622	55	742–750
24	494–502	40	622–630	56	750–758
25	505–510	41	630–638	57	758–766
26	510–518	42	638–646	58	766–774
27	518–526	43	646–654	59	774–782
28	526–534	44	654–662	60	782–790
29	534–542	45	662–670	61	790–798
30	542–550	46	670–678	62	798–806
31	550–558	47	678–686	63	806–814
32	558–566	48	686–694	64	814–822
33	566–574	49	694–702	65	822–830
34	574–582	50	702–710	66	830–838
35	582–590	51	710–718	67	838–846
36	590–598	52	718–726	68	846–854

Aerial dimenions

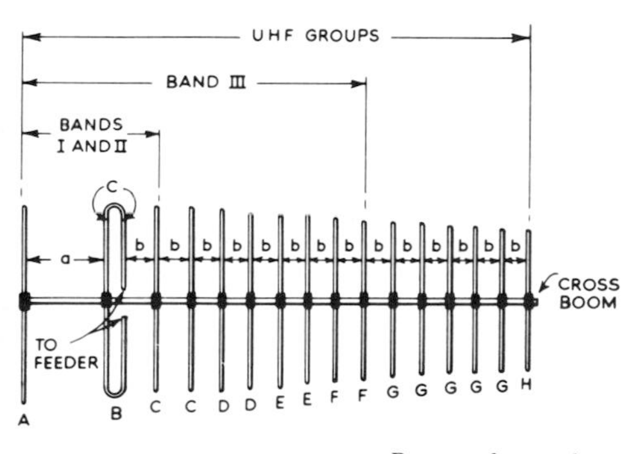

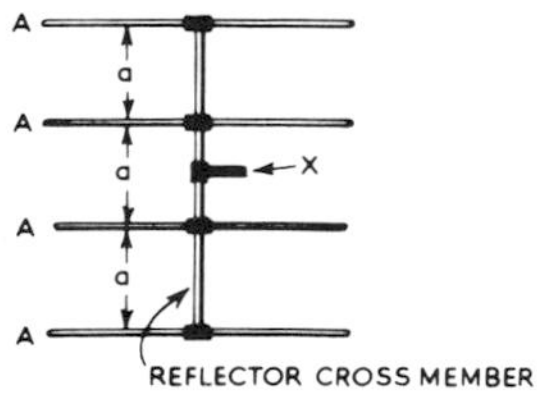

Pattern of general-purpose Yagi array to be used in conjunction with the dimensions given on p. 134.

Aerial dimensions

Channel	Dimensions in cm							
	A	B	C	D	E	F	G	H
UHF Groups								
A	30·1	30	24·1	23	22·8	21·1	20·4	19·9
B	26·5	21·7	18·9	18	17·8	16·5	16	15·5
C	23·2	18·2	16	15·3	15	14	13·3	12·2
D	26·1	23·5	18·4	16	15·5	14·8	13·8	13
E	27	26·5	21·1	18·6	17·9	17·6	16	15·8

UHF television channels and transmitters

Including transmitters to enter service during 1987

	ITV	Channel 4	BBC 1	BBC 2	Polarisation	Max vision erp(kW)
East of England						
Tacolneston	59	65	62	55	H	250
West Runton	23	29	33	26	V	2
Aldeburgh	23	30	33	26	V	10
Thetford	23	29	33	26	V	0·02
Little Walsingham	41	47	51	44	V	0·011
Creake	49	42	39	45	V	0·005
Wells next the sea	50	—	43	—	V	0·09
Burnham	46	—	40	—	V	0·077
Norwich (Central)	49	42	39	45	V	0·034
Bury St. Edmunds	25	32	22	28	V	0·017
Linnet Valley	23	29	33	26	V	0·016
Sudbury	41	47	51	44	H	250
Woodbridge	61	54	58	64	V	0·1
Ipswich (Stoke)	25	32	22	28	V	0·007
Wivenhoe Park	61	54	58	64	V	0·011
Felixstowe	60	67	31	63	V	0·005
Sandy Heath	24	21	31	27	H	1000
Northampton (Dall. Park)	56	68	66	62	V	0·065
Luton	59	65	55	62	V	0·08
Kings Lynn	52	—	48	—	V	0·34
The Borders and Isle of Man						
Caldbeck	28	32	30	34	H	500
Kendal	61	54	58	64	V	2
Windermere	41	47	51	44	V	0·5
Coniston	24	31	21	27	V	0·09
Hawkshead	23	29	33	26	V	0.061
Whitehaven	43	50	40	46	V	2
Keswick	24	31	21	27	V	0·12
Threlkeld	60	53	57	63	V	0·01
Ainstable	42	49	52	45	V	0·1
Haltwhistle	59	65	55	62	V	2
Gosforth	61	54	58	64	V	0·05
Bassenthwaite	49	42	52	45	V	0·16
Pooley Bridge	46	50	48	40	V	0·01
Moffat	42	49	52	45	V	0·00

a	*b*	*c*
10·3	10·3	1·8
8·9	8·9	1·8
7·5	7·5	1·8
7·6	7·6	1·8
15·8	15·8	1·8

Channels covered in the uhf groups are:

Group letter	*Colour code*	*Channels*
A	Red	21–34
B	Yellow	39–51
C	Green	50–66
D	Blue	49–68
E	Brown	39–68

	ITV	*Channel 4*	*BBC 1*	*BBC 2*	*Polarisation*	*Max vision erp(kW)*
Douglas	48	56	68	66	V	2
Beary Peark	43	50	40	46	V	0·25
Port St. Mary	61	54	58	64	V	0·25
Laxey	61	54	58	64	V	0·025
Langholm	60	53	57	63	V	0·025
Thornhill	60	53	57	63	V	0·5
Barskeoch Hill	59	65	55	62	V	2
New Galloway	23	29	33	26	V	0·1
Stranraer	60	53	57	63	V	0·25
Portpatrick	61	54	58	64	V	0·006
Cambret Hill	41	47	44	51	H	16
Creetown	61	54	58	64	V	0·032
Kirkcudbright	24	31	21	27	V	0·006
Glenluce	61	54	58	64	V	0·015
St. Bees	61	54	58	64	V	0·012
Workington	61	54	58	64	V	0·01
Bleachgreen	60	53	57	63	V	0·006
Dumfries South	46	50	40	48	V	0·023
Dentdale	60	53	57	63	V	0·052
Union Mills	52	42	39	45	V	0·012
Lowther Valley	46	50	48	40	V	0·026
Pinwherry	25	32	22	28	V	0·056
Ballantrae	61	54	58	64	V	0·0066
Lorton	60	53	57	63	V	0·05
Greystoke	60	53	57	63	V	0·011
Kirkby Stephen	60	53	57	63	V	0·012
Ravenstonedale	60	53	57	63	V	0·011
Orton	43	50	40	46	V	0·031
Sedbergh	43	50	40	46	V	0·5
Grasmere	60	53	57	63	V	0·02
Crosby Ravensworth	60	53	57	63	V	0·006
Crosthwaite	60	53	57	63	V	0·012
Selkirk	59	65	55	62	H	50
Eyemouth	23	29	33	26	V	2
Galashiels	41	47	51	44	V	0·1
Hawick	23	29	33	26	V	0·05
Jedburgh	41	47	51	44	V	0·16
Bonchester Bridge	49	42	39	45	V	0·006

	ITV	Channel 4	BBC 1	BBC 2	Polarisation	Max vision erp(kW)
Lauder	25	32	22	28	V	0·011
Peebles	25	32	22	28	V	0·1
Innerleithen	61	54	58	64	V	0·1
Berwick-upon-Tweed	24	31	21	27	V	0·038
Stow	23	29	33	26	V	0·005
Yetholm	41	47	51	44	V	0·006

Midlands (West)

	ITV	Channel 4	BBC 1	BBC 2	Polarisation	Max vision erp(kW)
Sutton Coldfield	43	50	46	40	H	1000
Kinver	56	68	66	48	H	0·012
Kidderminster	61	54	58	64	V	2
Brierley Hill	60	53	57	63	V	10
Bromsgrove	24	21	31	27	V	4
Malvern	66	68	56	62	V	2
Lark Stoke	23	29	33	26	V	7·6
Leek	25	32	22	28	V	1
Fenton	24	21	31	27	V	10
Hartington	56	68	66	48	V	0·033
Over Norton	55	67	65	48	V	0·031
Bretch Hill	55	67	65	48	V	0·087
Icomb Hill	25	32	22	28	V	0·11
Leamington Spa	66	68	56	62	V	0·2
Allesley Park	25	32	22	28	V	0·033
Cheadle	56	68	48	66	V	0·024
Tenbury Wells	60	53	57	63	V	0·014
Redditch	25	32	22	28	V	0·0016
Ironbridge	61	54	58	64	V	0·011
Guiting Power	41	47	51	44	V	0·012
Ipstones Edge	60	53	57	63	V	0·028
Whittingslow	60	53	57	63	V	0·056
Oakamoor	24	31	21	27	V	0·011
Turves Green	62	68	56	66	V	0·012
Brailes	34	59	30	52	V	0·04
Woodford Halse	25	32	22	28	V	0·007
Winshill	56	68	66	48	H	0·006
Winchcombe	61	54	58	64	V	0·006
Oxford	60	53	57	63	H	500
Charlbury	41	47	51	44	V	0·0033
Ascott-under-Wychwood	24	31	21	27	V	0·029
The Wrekin	23	29	26	33	H	100
Clun	59	65	55	62	V	0·056
Ridge Hill	25	32	22	28	H	100
Kington	49	42	39	45	V	0·025
Garth Hill	60	53	57	63	V	0·025
Ludlow	42	49	39	45	V	0·025
Hazler Hill	41	47	51	44	V	0·025
Oakeley Mynd	49	42	39	45	V	0·05
St. Briavels	43	50	40	46	V	0·012
Peterchurch	60	53	57	63	V	0·076
Andoversford	59	65	55	62	V	0·056
New Radnor	41	47	51	44	V	0·125
Hope-under-Dinmore	60	53	63	57	V	0·0018

	ITV	Channel 4	BBC 1	BBC 2	Polarisation	Max vision erp(kW)
Upper Soudley	43	50	40	46	V	0·0017
Eardiston	61	54	58	64	V	0·006
Midlands (East)						
Waltham	61	54	58	64	H	250
Ashbourne	25	32	22	28	V	0·25
Ambergate	25	32	22	28	V	0·03
Nottingham	24	31	21	27	V	2
Belper	68	62	66	56	V	0·03
Eastwood	23	29	33	26	V	0·0072
Stamford	49	42	39	45	V	0·0032
Parwich	24	31	21	27	V	0·0031
Stanton Moor	59	65	55	62	V	2
Bolehill	53	60	63	57	V	0·25
Matlock	24	31	21	27	V	0·017
Ashford-in-the-Water	23	29	33	26	V	0·011
Channel Islands						
Fremont Point	41	47	51	44	H	20
St. Helier	59	65	55	62	V	0·034
Les Touillets	54	52	56	48	H	2
Alderney	61	68	58	64	V	0·1
St. Peter Port	24	31	21	27	V	0·0014
Torteval	46	66	50	40	V	0·02
Gorey	23	29	54	26	V	0·006
Lancashire						
Winter Hill	59	65	55	62	H	500
Darwen	49	42	39	45	V	0·5
Pendle Forest	25	32	22	28	V	0·5
Haslingden	23	29	33	26	V	8
Elton	24	31	21	27	V	0·063
Saddleworth	49	42	52	45	V	0·5
Storeton	25	32	22	28	V	2·8
Bacup	43	53	40	46	V	0·25
Ladder Hill	23	29	33	26	V	1
Bidston	30	47	51	44	V	0·066
Birch Vale	43	53	40	46	V	0·25
Whitworth	25	32	22	28	V	0·05
Glossop	25	32	22	28	V	1
Buxton	24	31	21	27	V	1
Trawden	60	67	57	63	V	0·2
Whalley	43	53	40	46	V	0·05
Littleborough	24	31	21	27	V	0·5
North Oldham	24	31	21	27	V	0·04
Macclesfield	25	32	22	28	V	0·037
Congleton	41	47	51	44	V	0·2
Oakenhead	41	47	51	44	V	0·1
Whitewell	60	67	57	63	V	0·08
Delph	23	29	33	26	V	0·003
Lancaster	24	21	31	27	V	10
Blackburn	41	47	51	44	V	0·008

	ITV	Channel 4	BBC 1	BBC 2	Polarisation	Max vision erp(kW)
Millom Park	25	32	22	28	V	0·25
Ramsbottom	56	68	48	66	V	0·08
Dalton	43	53	40	46	V	0·025
Over Biddulph	30	48	34	67	V	0·022
Haughton Green	43	53	40	46	H	0·007
Parbold	41	47	51	44	V	0·036
Chinley	61	67	57	64	V	0·012
Dog Hill	43	53	40	46	V	0·085
Romiley	41	47	51	44	V	0·011
Bollington	24	31	21	27	V	0·021
Langley	24	31	21	27	V	0·0045
Ribblesdale	41	47	51	44	V	0·03
Backbarrow	60	50	57	63	V	0·003
West Kirby	24	31	34	27	V	0·013
Brook Bottom	61	68	58	64	V	0·006
Staveley-in-Cartmel	43	53	40	46	V	0·01
Penny Bridge	23	29	33	26	V	0·031
Cartmel	25	32	22	28	H	0·0022
Urswick	41	47	51	44	V	0·0058
Melling	60	53	57	63	V	0·025
Austwick	49	42	39	45	V	0·032
Chatburn	23	29	33	26	V	0·007
Woodnook	49	52	39	45	V	0·003
Middleton	30	48	67	34	V	0·006
Wardle	25	32	22	28	H	0·003
Norden	30	57	34	67	V	0·009
Brinscall	24	31	27	21	V	0·0008
North-east Scotland						
Durris	25	32	22	28	H	500
Peterhead	59	65	55	62	V	0·1
Gartly Moor	61	54	58	64	V	2·2
Rosehearty	41	47	51	44	V	2
Balgownie	43	50	40	46	V	0·04
Tullich	59	65	55	62	V	0·07
Braemar	42	49	39	45	V	0·015
Tomintoul	43	50	40	46	V	0·0065
Banff	42	49	39	45	V	0·028
Ellon	49	42	39	45	V	0·0027
Brechin	43	50	40	46	V	0·0065
Boddam	42	49	39	45	V	0·006
Angus	60	53	57	63	H	100
Perth	49	42	39	45	V	1
Crieff	23	29	33	26	V	0·1
Cupar	41	47	51	44	V	0·02
Pitlochry	25	32	22	28	V	0·15
Kenmore	23	29	33	26	V	0·12
Blair Atholl	43	50	40	46	V	0·05
Tay Bridge	41	47	51	44	V	0·5
Killin	49	42	39	45	V	0·13
Auchtermuchty	49	42	39	45	V	0·05
Camperdown	23	29	33	26	V	0·002

	ITV	Channel 4	BBC 1	BBC 2	Polarisation	Max vision erp(kW)
Strathallan	49	42	39	45	V	0·029
Methven	25	32	22	28	V	0·006
Dunkeld	41	47	51	44	V	0·1
Keelylang Hill (Orkney)	43	50	40	46	H	100
Pierowall	23	29	33	26	V	0·0072
Bressay	25	32	22	28	V	10
Fitful Head	49	42	39	45	V	0·094
Scalloway	59	65	55	62	V	0·029
Swinister	59	65	55	62	V	0·21
Baltasound	42	49	39	45	V	0·018
Fetlar	43	50	40	46	V	0·013
Collafirth Hill	41	47	51	44	V	0·41
Weisdale	61	54	58	64	V	0·06
Burgar Hill	24	31	21	27	V	0·0055
Rumster Forest	24	21	31	27	H	100
Ben Tongue	49	42	39	45	V	0·04
Thurso	60	53	57	63	V	0·0027
Melvich	41	47	51	44	V	0·055
Durness	53	60	57	63	V	0·007
Knock More	23	29	33	26	H	100
Grantown	41	47	51	44	V	0·35
Kingussie	43	50	40	46	V	0·091
Craigellachie	60	53	57	63	V	0·07
Balblair Wood	59	65	55	62	V	0·083
Lairg	41	47	51	44	V	0·013
Avoch	53	60	63	57	V	0·004
Eitshal (Lewis)	23	29	33	26	H	100
Scoval	59	65	55	62	V	0·16
Clettraval	41	47	51	44	V	2
Daliburgh (South Uist)	60	53	57	63	V	0·03
Skriaig	24	31	21	27	V	1
Penifiler	49	42	39	45	V	0·04
Duncraig	41	47	51	44	V	0·16
Attadale	25	32	22	28	V	0·0088
Badachro	43	50	40	46	V	0·035
Ness of Lewis	41	47	51	44	V	0·032
Ullapool	49	52	39	45	V	0·078
Kilbride (South Uist)	49	42	39	45	V	0·13
Uig	43	50	53	46	V	0·0033
Ardintoul	49	42	39	45	V	0·047
Tarbert (Harris)	49	52	39	45	V	0·047
Bruernish	43	50	40	46	V	0·0069
Poolewe	47	41	51	44	V	0·02
Lochinver	43	50	40	46	V	0·008
Rosemarkie	49	42	39	45	H	100
Auchmore Wood	25	32	22	28	V	0·1
Fort Augustus	23	29	33	26	V	0·011
Fodderty	60	53	57	63	V	0·12
Wester Erchite	24	31	21	27	V	0·016
Glen Urquhart	41	47	51	44	V	0·09
Tomatin	25	32	22	28	V	0·012
Inverness	65	59	55	62	V	0·05
Tomich	24	31	21	27	V	0·014

	ITV	Channel 4	BBC 1	BBC 2	Polarisation	Max vision erp(kW)
Wales						
Wenvoe	41	47	44	51	H	500
Kilvey Hill	23	29	33	26	V	10
Rhondda	23	29	33	26	V	2·5
Mynydd Machen	23	29	33	26	V	2
Maesteg	25	32	22	28	V	0·25
Pontypridd	25	32	22	28	V	0·5
Aberdare	24	31	21	27	V	0·5
Merthyr Tydfil	25	32	22	28	V	0·13
Bargoed	24	31	21	27	V	0·3
Rhymney	60	53	57	63	V	0·15
Clydach	23	29	33	26	V	0·0035
Abertillery	25	32	22	28	V	0·28
Ebbw Vale	59	65	55	62	V	0·5
Blaina	43	50	40	46	V	0·1
Pontypool	24	31	21	27	V	0·25
Cilfrew	49	52	39	45	V	0·015
Blaenavon	60	53	57	63	V	0·15
Abergavenny	49	42	39	45	V	1
Ferndale	60	53	57	63	V	0·08
Porth	43	50	40	46	V	0·08
Wattsville	60	53	63	57	V	0·0052
Llangeinor	59	65	55	62	V	0·19
Treharris	52	68	56	48	V	0·05
Cwmafon	24	31	21	27	V	0·07
Llyswen	24	31	21	27	V	0·03
Llanhilleth	49	42	39	45	V	0·03
Gilfach Goch	24	31	21	27	V	0·05
Taff's Well	59	65	55	62	V	0·052
Ogmore Vale	60	53	57	63	V	0·1
Abertridwr	60	53	57	63	V	0·05
Ynys Owen	59	65	55	62	V	0·08
Tonypandy	59	65	55	62	V	0·02
Fernhill	59	65	62	55	V	0·0031
Mynydd Bach	61	54	58	64	V	0·25
Bedlinog	24	31	21	27	V	0·01
Machen Upper	62	68	55	65	V	0·009
Cwm Ffrwd-Oer	43	50	39	46	V	0·003
Blaenau-Gwent	60	53	57	63	V	0·0028
Pennar	43	50	40	46	V	0·1
Brecon	61	54	58	64	V	1
Sennybridge	43	50	40	46	V	0·064
Clyro	41	47	51	44	V	0·16
Crickhowell	24	31	21	27	V	0·15
Blackmill	25	32	22	28	V	0·01
Pennorth	23	29	33	26	V	0·05
Pontardawe	61	68	58	64	V	0·13
Deri	25	32	22	28	V	0·05
Cwmaman	49	42	39	45	V	0·0014
Ton Pentre	61	54	58	64	V	0·08
Trecastle	25	32	22	28	V	0·00
Monmouth	59	65	55	62	V	0·05

	ITV	Channel 4	BBC 1	BBC 2	Polarisation	Max vision erp(kW)
Cwmfelinfach	48	42	52	45	V	0·006
Llanfoist	60	53	57	63	V	0·018
Abercynon	58	54	64	66	H	0·0062
Tynewydd	59	65	55	62	V	0·02
Craig-Cefn-Parc	43	50	46	40	V	0·0063
Briton Ferry	43	50	46	40	V	0·02
Dowlais	61	54	58	64	V	0·013
Rhondda Fach	25	32	22	28	V	0·0015
Trefechan (Merthyr)	42	49	39	45	V	0·005
Crucorney	24	31	21	27	V	0·011
Tonyrefail	59	65	55	62	V	0·02
Efail Fach	49	52	39	45	V	0·0084
Llanharan	24	31	21	27	V	0·0017
Burry Port	61	54	58	64	V	0·0031
Rhondda 'B'	49	68	66	39	H	0·005
Gelli-Fendigaid	59	65	55	62	H	0·012
South Maesteg	59	65	55	62	V	0·0059
Llanddona	60	53	57	63	H	100
Betws-y-Coed	24	31	21	27	V	0·5
Penmaen Rhos	25	32	22	28	H	0·141
Conway	43	50	40	46	V	2
Bethesda	60	53	57	63	V	0·025
Deiniolen	25	32	22	28	V	0·05
Arfon	41	47	51	44	V	3·6
Llandecwyn	61	54	58	64	V	0·3
Ffestiniog	25	32	22	28	V	1·2
Waunfawr	25	32	22	28	V	0·026
Amlwch	25	32	22	28	V	0·035
Cemaes	43	50	40	46	V	0·012
Mochdre	23	29	33	26	V	0·0017
Dolwyddelan	41	47	51	44	V	0·011
Llanengan	61	54	58	64	H	0·003
Carmel	60	53	57	63	H	100
Llanelli	49	67	39	45	V	0·1
Ystalyfera	49	42	39	45	V	0·05
Llandrindod Wells	49	42	39	45	V	2·25
Rhayader	23	29	33	26	V	0·1
Llanwrtyd Wells	24	31	21	27	V	0·01
Builth Wells	25	32	22	28	V	0·026
Tenby	49	42	39	45	V	0·032
Cwmgors	24	31	21	27	V	0·026
Abercraf	25	32	22	28	V	0·13
Mynydd Emroch	43	50	40	46	V	0·09
Greenhill	24	31	21	27	V	0·074
Penderyn	49	42	39	45	V	0·012
Talley	49	42	39	45	V	0·0065
Llansawel	32	25	22	28	V	0·0065
Presely	43	50	46	40	H	100
Mynydd Pencarreg	61	54	58	64	V	0·12
Tregaron	56	66	62	68	V	0·015
Llandyfriog	25	32	22	28	V	0·11
St. Dogmaels	23	29	33	26	V	0·015

	ITV	Channel 4	BBC 1	BBC 2	Polarisation	Max vision erp(kW)
Trefin	25	32	22	28	V	0·056
Abergwynfi	24	31	21	27	V	0·003
Glyncorrwg	49	42	39	45	V	0·0007
Llwyn Onn	25	32	22	28	V	0·05
Dolgellau	59	65	55	62	V	0·03
Croeserw	61	54	58	64	V	0·12
Pencader	23	29	33	26	V	0·006
Llandysul	60	53	57	63	V	0·05
Broad Haven	61	54	58	64	V	0·006
Rheola	59	65	55	62	V	0·1
Newport Bay	60	67	57	63	V	0·013
Ferryside	24	31	21	27	V	0·007
Llangybi	25	32	22	28	V	0·012
Blaen-Plwyf	24	21	31	27	H	100
Machynlleth	60	53	57	63	V	0·02
Aberystwyth	61	54	58	64	V	0·023
Fishguard	61	54	58	64	V	0·056
Long Mountain	61	54	58	64	V	1
Llandinam	41	47	44	51	V	0·25
Llanidloes	25	32	22	28	V	0·005
Llanfyllin	25	32	22	28	V	0·13
Moel-y-Sant	24	31	34	27	V	0·11
Kerry	24	31	21	27	V	0·017
Carno	24	31	21	27	V	0·011
Dolybont	61	54	58	64	V	0·032
Llanbrynmair	25	32	22	28	V	0·02
Afon Dyfi	25	32	22	28	V	0·0063
Llangurig	23	29	33	26	V	0·008
Trefilan	60	53	57	63	V	0·086
Llanrhaedr-ym-Mochnant	49	42	39	45	V	0·077
Bow Street	41	47	51	44	V	0·02
Ynys-Pennal	41	47	51	44	V	0·02
Llangadfan	25	32	22	28	V	0·0063
Tregynon	25	32	22	28	V	0·035
Corris	49	42	39	45	V	0·006
Llangynog	65	59	55	62	V	0·006
Moel-y-Parc	49	42	52	45	H	100
Llangollen	60	53	57	63	V	0·015
Glyn Ceiriog	61	54	58	64	V	0·007
Bala	23	29	33	26	V	0·2
Corwen	25	32	22	28	V	0·3
Pontfadog	25	32	22	28	V	0·0064
Cerrigydrudion	23	29	33	26	V	0·032
Wrexham-Rhos	—	67	39	—	V	0·2
Llanuwchllyn	43	50	40	46	V	0·03
Cefn-Mawr	41	47	51	44	V	0·034
Llanarmon-yn-ial	24	31	21	27	V	0·006
Llangernyw	32	25	22	28	V	0·007
Betws-yn-Rhos	24	31	21	27	V	0·013
Glyndyfrdwy	59	65	55	62	V	0·0056
Llandderfel	65	59	55	62	V	0·0065
Llanddulas	23	29	33	26	H	0·012
Pwll-Glas	23	29	33	26	V	0·007

West of England	ITV	Channel 4	BBC 1	BBC 2	Polarisation	Max vision erp(kW)
Mendip	61	54	58	64	H	500
Crockerton	41	47	51	44	V	0·077
Bath	25	32	22	28	V	0·25
Westwood	43	50	40	46	V	0·1
Avening	41	47	51	44	V	0·0056
Calne	24	31	21	27	V	0·05
Redcliff Bay	34	67	30	56	H	0·011
Bristol KWH	42	52	45	48	V	1
Bristol IC	43	50	40	46	V	0·5
Washford	39	68	49	66	V	0·062
Easter Compton	34	67	30	56	V	0·01
West Lavington	24	31	21	27	V	0·012
Seagry Court (Swindon)	41	47	44	51	V	0·0025
Coleford	45	39	42	52	V	0·01
Monksilver	52	42	45	48	V	0·015
Ogbourne St. George	43	50	40	46	V	0·013
Wootton Courtenay	25	32	22	28	V	0·056
Stroud	42	52	48	45	V	0·5
Cirencester	23	29	33	26	V	0·25
Nailsworth	23	29	33	26	V	0·031
Chalford	24	31	21	27	V	0·13
Roadwater	24	31	21	27	H	0·012
Marlborough	25	32	22	28	V	0·1
Upavon	23	29	33	26	V	0·07
Porlock	42	52	48	45	V	0·025
Countlsbury	49	67	39	56	H	0·11
Cerne Abbas	25	32	22	28	V	0·11
Hutton	39	68	49	66	V	0·14
Bristol (Montpelier)	23	29	33	26	V	0·01
Box	43	50	40	46	V	0·0068
Dursley (Uley)	43	50	40	46	V	0·055
Slad	23	29	33	26	H	0·0028
Frome	24	31	21	27	V	0·0018
Bristol (Barton House)	24	31	21	27	H	0·011
Bruton	43	50	40	46	V	0·0015
Kewstoke	34	67	30	56	V	0·012
Burrington	59	65	55	62	H	0·103
Ubley	24	31	21	27	V	0·079
Portishead	49	68	66	39	V	0·007
Backwell	25	32	22	28	V	0·094
Tintern	24	31	21	27	V	0·006
Chiseldon	34	67	30	49	V	0·02
Chepstow	24	31	21	27	V	0·0031
Blakeney	24	31	21	27	V	0·007
Lydbrook	43	50	40	46	V	0·0075
Parkend	41	47	51	44	V	0·0017
Clearwell	68	56	66	48	V	0·01
Woodcombe	24	31	21	27	V	0·0063
Exford	41	47	51	44	V	0·008
Kilve	39	68	49	66	H	0·008
Crewkerne	43	50	40	46	V	0·0016
Carhampton	30	56	34	67	V	0·008

	ITV	Channel 4	BBC 1	BBC 2	Polarisation	Max vision erp(kW)
London						
Crystal Palace	23	30	26	33	H	1000
Guildford	43	50	40	46	V	10
Hertford	61	54	58	64	V	2
Reigate	60	53	57	63	V	10
Hemel Hempstead	41	47	51	44	V	10
Woolwich	60	67	57	63	V	0·63
High Wycombe	59	65	55	62	V	0·5
Wooburn	56	68	49	52	V	0·1
Henley-on-Thames	67	54	48	64	V	0·1
Bishops Stortford	59	49	55	62	V	0·029
Chesham	43	50	40	46	V	0·1
Welwyn	43	50	40	46	V	0·15
Gt. Missenden	61	54	58	64	V	0·085
Mickleham	58	68	61	55	V	0·09
Kenley	43	50	40	46	V	0·14
Chepping Wycombe	41	47	51	44	V	0·02
Hughenden	43	50	40	46	V	0·06
Forest Row	62	66	48	54	V	0·12
Chingford	52	48	56	50	V	0·0075
Hemel Hempstead (Town)	61	54	58	64	V	0·013
Walthamstow North	49	68	45	66	V	0·0017
Marlow Bottom	61	54	58	64	V	0·011
Cane Hill	58	68	61	54	V	0·018
New Addington	54	68	64	48	V	0·017
West Wycombe	43	67	40	46	V	0·028
Otford	60	53	57	63	V	0·031
Lea Bridge	39	59	55	62	V	0·006
Micklefield	57	67	54	64	V	0·0062
Alexandra Palace	61	54	58	64	H	0·065
Dorking	41	47	51	44	H	0·04
Caterham	59	65	55	62	V	0·035
East Grinstead	46	59	40	56	V	0·117
Biggin Hill	49	67	45	52	V	0·008
Croydon (Old Town)	52	67	49	56	V	0·033
Skirmett	41	47	51	44	V	0·13
St. Albans	57	67	49	63	V	0·022
Gravesend	59	49	55	62	V	0·011
Wonersh	52	67	48	65	V	0·025
New Barnet	59	48	55	62	V	0·007
Hammersmith	59	65	48	62	V	0·01
Central Scotland						
Black Hill	43	50	40	46	H	500
Kilmacolm	24	31	21	27	V	0·032
South Knapdale	60	53	57	63	V	1·45
Biggar	25	32	22	28	V	0·5
Abington	60	53	57	63	H	0·0051
Glasgow WC	56	66	68	62	V	0·032
Killearn	59	55	65	62	V	0·5
Callander	25	32	22	28	V	0·1
Cathcart	60	53	57	63	V	0·002

	ITV	Channel 4	BBC 1	BBC 2	Polarisation	Max vision erp(kW)
Torosay	25	32	22	28	V	20
Cow Hill	43	50	40	46	V	0·065
Netherton Braes	25	32	22	28	V	0·005
Gigha Island	41	47	51	44	V	0·06
Tarbert (Loch Fyne)	24	31	21	27	V	0·0036
Glengorm	48	54	56	52	V	1·1
Mallaig	40	50	43	46	V	0·018
Ballachulish	23	29	33	26	V	0·018
Haddington	61	54	58	64	V	0·02
Kinlochleven	59	65	55	62	V	0·012
Onich	61	54	58	64	V	0·017
Strachur	23	29	33	26	V	0·035
Spean Bridge	24	31	21	27	V	0·07
Oban	41	47	51	44	V	0·012
Bellanoch	42	49	39	45	V	0·05
Castlebay	24	31	21	27	V	0·0066
Dalmally	41	47	51	44	V	0·041
Dollar	61	54	58	64	V	0·01
Ravenscraig	24	31	21	27	V	0·02
Kirkfieldbank	60	53	57	63	V	0·0058
Tillicoultry	60	53	57	63	V	0·005
Fintry	24	31	34	27	V	0·019
Fiunary	43	50	40	46	V	0·05
Twechar	25	32	22	28	V	0·007
Strathblane	24	31	21	27	V	0·0071
Broughton	24	31	21	27	V	0·007
Leadhills	61	54	58	64	V	0·003
Glespin	61	54	58	64	V	0·006
Craigkelly	24	21	31	27	H	100
Penicuik	61	54	58	64	V	2
West Linton	23	29	33	26	V	0·025
Aberfoyle	61	54	58	64	V	0·087
Darvel	23	29	33	26	H	100
Muirkirk	41	47	51	44	V	0·1
Kirkconnel	61	54	58	64	V	0·25
West Kilbride	41	47	51	44	V	0·35
Lethanhill	60	53	57	63	V	0·25
Girvan	59	65	55	62	V	0·25
Campbeltown	60	53	57	63	V	0·13
Port Ellen	25	32	22	28	V	0·09
Bowmore	49	42	39	45	V	0·08
Millburn Muir	42	49	39	52	V	0·25
Rosneath	61	54	58	64	V	10
Rosneath	61	54	58	64	H	0·05
Millport	61	54	58	64	H	0·0027
Troon	61	54	58	64	V	0·02
Rothesay	25	32	22	28	V	2
Tighnabruaich	49	42	39	45	V	0·092
Lochwinnoch	60	53	57	63	H	0·086
New Cumnock	43	50	40	46	V	0·012
Rothesay Town	59	65	55	62	V	0·0054
Claonaig	59	65	55	62	V	0·074

	ITV	Channel 4	BBC 1	BBC 2	Polarisation	Max vision erp(kW)
Carradale	41	47	51	44	V	0·029
Ardentinny	49	52	39	45	V	0·07
Arrochar	24	31	21	27	V	0·006
Ardnadam	41	47	51	44	V	0·017
Garelochhead	41	47	51	44	V	0·012
Wanlockhead	47	41	51	44	V	0·002
Kirkoswald	25	32	22	28	V	0·032
Kirkmichael	49	52	39	45	V	0·019
Dunure	43	50	40	46	V	0·012
Holmhead	41	47	51	44	V	0·012
Largs	42	49	39	45	H	0·012
Sorn	43	50	40	46	V	0·0065
South of England						
Rowridge	27	21	31	24	H	500
Salisbury	60	53	57	63	V	10
Till Valley	43	50	46	40	V	0·075
Ventnor	49	42	39	45	V	2
Poole	60	53	57	63	V	0·1
Brighton	60	53	57	63	V	10
Shrewton	41	47	51	44	V	0·0045
Findon	41	47	51	44	V	0·05
Patcham	43	50	46	40	H	0·069
Winterborne Stickland	43	50	40	46	V	1
Corfe Castle	41	47	51	44	V	0·014
Portslade	41	47	51	44	V	0·019
Westbourne	41	47	51	44	V	0·038
Ovingdean	44	68	65	42	V	0·019
Saltdean	55	47	51	66	V	0·014
Donhead	41	47	51	44	V	0·029
Millbrook	41	47	51	44	V	0·035
Brighstone	41	47	51	44	V	0·14
Hangleton	49	42	39	45	V	0·0068
Lulworth	59	65	55	62	V	0·011
Piddletrenthide	49	42	39	45	V	0·056
Winterbourne Steepleton	45	66	39	49	V	0·012
Cheselbourne	53	60	57	63	V	0·0065
Brading	41	47	51	44	V	0·004
Midhurst	58	68	61	55	H	100
Haslemere	25	32	22	28	V	0·015
Hannington	42	66	39	45	H	250
Tidworth	32	25	22	28	V	0·01
Chisbury	59	52	55	62	V	0·025
Sutton Row	25	32	22	28	V	0·25
Alton	59	52	49	62	V	0·01
Hemdean (Caversham)	56	59	49	52	V	0·022
Aldbourne	24	31	21	27	V	0·007
Lambourn	59	52	55	62	V	0·007
Luccombe (IOW)	59	34	56	62	V	0·025
Dover	66	53	50	56	H	100
Dover Town	23	30	33	26	V	0·1
Hythe	24	31	21	27	V	0·051

	ITV	Channel 4	BBC 1	BBC 2	Polarisation	Max vision erp(kW)
Chartham	24	31	21	27	V	0·1
Faversham	25	32	22	28	V	0·013
Rye	41	47	58	44	V	0·012
Newnham	24	31	21	27	V	0·035
Lyminge	25	32	22	28	V	0·0069
Horn Street	41	47	58	44	V	0·003
Elham	23	30	33	26	V	0·0035
Heathfield	64	67	49	52	H	100
Tunbridge Wells	41	47	51	44	V	10
St. Marks	60	53	57	63	V	0·051
Newhaven	43	41	39	45	V	2
Hastings	28	32	22	25	V	1
Eastbourne	23	30	33	26	V	0·125
Haywards Heath	43	41	39	45	V	0·037
Wye (Ashford)	25	32	22	28	V	0·031
East Dean	54	42	62	44	V	0·008
Hamstreet	23	30	33	26	V	0·0007
Lamberhurst	62	58	54	60	V	0·003
Mountfield	24	31	21	27	V	0·0035
Sedelscombe	23	30	33	26	V	0·007
Steyning	62	56	45	59	V	0·14
Bluebell Hill	43	65	40	46	H	30
Chatham Town	61	54	58	68	V	0·011
North-east England						
Pontop Pike	61	54	58	64	H	500
Newton	23	29	33	26	V	2
Fenham	24	31	21	27	V	2
Weardale	41	47	44	51	V	1
Alston	49	42	52	45	V	0·4
Catton Beacon	43	50	40	46	V	0·14
Morpeth	25	32	22	28	V	0·044
Bellingham	24	31	21	27	V	0·05
Humshaugh	49	42	39	45	V	0·059
Haydon Bridge	41	47	51	44	V	0·1
Shotley Field	25	32	22	28	V	0·2
Durham	43	50	40	46	V	0·015
Ireshopeburn	59	65	55	62	V	0·011
Hedleyhope	43	50	40	46	H	0·018
Seaham	41	47	51	44	V	0·059
Sunderland	43	50	40	46	V	0·013
Staithes	41	47	51	44	V	0·0017
Esh	49	42	39	45	V	0·012
Falstone	41	47	51	44	V	0·0063
Bilsdale	29	23	33	26	H	500
Whitby	59	65	55	62	V	0·25
Bainbridge	60	53	57	63	V	0·031
Grinton Lodge	43	50	40	46	V	0·025
Guisborough	60	53	57	63	V	0·05
Ravenscar	61	54	58	64	V	0·2
Limber Hill	43	50	40	46	V	0·05
Skinningrove	43	50	40	46	V	0·014

	ITV	Channel 4	BBC 1	BBC 2	Polarisation	Max vision erp(kW)
Romaldkirk	41	47	51	44	V	0·058
West Burton	43	50	40	46	V	0·013
Aislaby	52	49	39	45	V	0·04
Rosedale Abbey	43	50	40	46	V	0·007
Peterlee (Horden)	49	39	45	52	V	0·002
Chatton	49	42	39	45	H	100
Rothbury	65	59	55	62	V	0·05
Northern Ireland						
Divis	24	21	31	27	H	500
Larne	49	42	39	45	V	0·5
Carnmoney Hill	43	50	40	46	V	0·1
Kilkeel	49	42	39	45	V	0·5
Newcastle	59	65	55	62	V	1
Armagh	49	42	39	45	V	0·12
Black Mountain	49	42	39	45	V	0·025
Whitehead	52	67	48	56	V	0·012
Bellair	52	67	48	56	V	0·04
Draperstown	49	42	39	45	V	0·0118
Moneymore	49	42	39	45	V	0·0067
Newry North	41	47	51	44	V	0·01
Rostrevor Forest	46	50	48	40	V	0·058
Newry South	49	42	39	45	V	0·02
Benagh	25	32	22	28	V	0·056
Cushendun	32	25	22	28	V	0·026
Cushendall	43	50	40	46	V	0·013
Glynn	61	54	58	64	V	0·0014
Newtownards	61	54	58	64	V	0·011
Banbridge	46	50	44	48	V	0·0061
Glenariff	61	54	58	64	V	0·011
Killowen Mountain	24	21	31	27	V	0·015
Bangor	59	65	62	55	V	0·003
Limavady	59	65	55	62	H	100
Londonderry	41	47	51	44	V	10
Ballycastle Forest	49	42	39	45	V	0·012
Bushmills	41	47	51	44	V	0·0065
Strabane	49	42	39	45	V	2
Claudy	60	53	57	63	V	0·029
Gortnalee	24	31	21	27	V	0·032
Castlederg	65	59	55	62	V	0·011
Plumbridge	56	68	52	66	V	0·0125
Glenelly Valley	23	29	33	26	V	0·012
Ballintoy	49	42	39	45	V	0·0017
Buckna	41	47	51	44	V	0·013
Gortnageeragh	42	49	39	45	V	0·019
Muldonagh	32	25	22	28	V	0·012
Brougher Mountain	25	32	22	28	H	100
Belcoo	41	47	51	44	V	0·087
Derrygonnelly	47	66	51	44	V	0·006
Lisbellaw	59	65	55	62	V	0·0065

South-west England	ITV	Channel 4	BBC 1	BBC 2	Polarisation	Max vision erp(kW)
Caradon Hill	25	32	22	28	H	500
St. Austell	59	65	55	62	V	0·1
Looe	43	50	40	46	V	0·005
Hartland	52	66	48	56	V	0·029
Gunnislake	43	50	40	46	V	0·04
Plympton (Plymouth)	61	54	58	64	V	2
Downderry	59	65	55	62	V	0·026
Tavistock	60	53	57	63	V	0·1
Woolacombe	42	49	39	45	V	0·006
Penaligon Downs	49	42	39	45	V	0·1
Newton Ferrers	59	65	55	62	V	0·0065
Ilfracombe	61	54	58	64	V	0·25
Combe Martin	49	42	39	45	V	0·1
Okehampton	49	42	39	45	V	0·1
Ivybridge	42	49	39	45	V	0·5
Kingsbridge	43	50	40	46	V	0·2
Penryn	59	65	55	62	V	0·022
Plymouth (North Road)	43	50	40	46	V	0·012
Slapton	55	68	48	66	V	0·125
Truro	61	54	58	64	V	0·022
Croyde	41	47	51	44	V	0·0015
Chambercombe	24	31	21	27	V	0·007
Salcombe	44	30	51	41	V	0·017
Polperro	60	53	57	63	V	0·0028
Mevagissey	43	50	40	46	H	0·0066
Lostwithiel	43	50	40	46	V	0·0063
Aveton Gifford	66	47	51	44	V	0·0015
Stockland Hill	23	29	33	26	H	250
St. Thomas (Exeter)	41	47	51	44	V	0·25
Beer	59	65	55	62	V	0·0029
Tiverton	43	50	40	46	V	0·1
Bampton	45	52	39	49	V	0·03
Culm Valley	49	42	39	45	V	0·058
Bridport	41	47	51	44	V	0·1
Beaminster	59	65	55	62	V	0·02
Weymouth	43	50	40	46	V	2
Dawlish	59	65	55	62	V	0·0066
Stokeinteignhead	41	47	51	44	V	0·0063
Dunsford	39	49	45	67	V	0·006
Crediton	43	50	40	46	V	0·04
Beacon Hill	60	53	57	63	H	100
Dartmouth	41	47	51	44	V	0·01
Ashburton	24	31	21	27	V	0·003
Teignmouth	45	67	39	49	V	0·025
Coombe	24	31	21	27	V	0·0065
Newton Abbot	43	50	40	46	V	0·003
Buckfastleigh	41	47	51	44	V	0·0062
Totnes	24	31	21	27	V	0·0034
Harbertonford	49	42	39	45	H	0·0018

	ITV	Channel 4	BBC 1	BBC 2	Polarisation	Max vision erp(kW)
Sidmouth	45	67	39	49	V	0·012
Occombe Valley	24	31	21	27	V	0·0008
Torquay Town	41	47	51	44	V	0·04
Hele	43	50	40	46	H	0·006
Edginswell	45	67	39	49	V	0·004
Huntshaw Cross	59	65	55	62	H	100
Swimbridge	23	29	33	26	V	0·0066
Westward Ho	24	31	21	27	V	0·032
Chagford	24	31	21	27	V	0·012
Brushford	24	31	21	27	V	0·02
North Bovey	43	50	40	46	V	0·034
Redruth	41	47	51	44	H	100
Isles of Scilly	24	31	21	27	V	0·5
St. Just	61	54	58	64	V	0·25
Helston	61	54	58	64	V	0·01
Bossiney	61	54	58	64	V	0·0074
Boscastle	23	29	33	26	V	0·0056
Portreath	23	29	33	26	V	0·0016
Praa Sands	59	65	55	62	V	0·01
Porthleven	23	29	33	26	H	0·0016
St. Anthony-in-Roseland	23	29	33	26	V	0·0017
Gulval	23	29	33	26	V	0·26
Yorkshire						
Emley Moor	47	41	44	51	H	1000
Wharfedale	25	32	22	28	V	2
Sheffield	24	21	31	27	V	5
Skipton	49	42	39	45	V	10
Chesterfield	23	29	33	26	V	2
Halifax	24	31	21	27	V	0·5
Keighley	61	54	58	64	V	10
Shatton Edge	48	54	52	58	V	1
Hebden Bridge	25	32	22	28	V	0·25
Ripponden	61	54	58	64	V	0·06
Cop Hill	25	32	22	28	V	1
Idle	24	31	21	27	V	0·25
Headingley	61	54	58	64	H	0·011

	ITV	Channel 4	BBC 1	BBC 2	Polarisation	Max vision erp(kW)
Beecroft Hill	59	65	55	62	V	1
Oxenhope	25	32	22	28	V	0·2
Calver Peak	49	42	39	45	V	0·25
Tideswell Moor	60	66	56	63	V	0·25
Hope	25	32	22	28	V	0·012
Addingham	43	50	40	46	V	0·025
Luddenden	60	67	57	63	V	0·059
Dronfield	59	65	55	62	H	0·003
Hasland	60	53	57	63	V	0·0065
Edale	60	53	57	63	V	0·004
Totley Rise	49	42	39	45	V	0·012
Cullingworth	49	68	66	39	H	0·013
Skipton Town	24	31	21	27	V	0·013
Batley	60	67	57	63	V	0·013
Heyshaw	60	53	57	63	V	0·5
Primrose Hill	60	67	57	63	V	0·028
Armitage Bridge	61	54	58	64	V	0·0065
Wincobank	59	65	55	62	V	0·0015
Holmfirth	56	68	49	66	V	0·026
Hagg Wood	59	65	55	62	V	0·033
Keighley Town	23	29	33	26	V	0·006
Sutton-in-Craven	23	29	33	26	V	0·012
Cragg Vale	61	54	58	64	V	0·025
Stocksbridge	61	54	58	64	V	0·012
Oughtibridge	59	65	55	62	V	0·02
Holmfield	59	65	55	62	V	0·022
Grassington	23	29	33	26	V	0·06
Cornholme	61	54	58	64	V	0·042
Walsden	60	67	57	63	V	0·05
Todmorden	49	42	39	45	V	0·5
Walsden South	43	53	40	46	V	0·006
Copley	59	65	55	62	V	0·0014
Kettlewell	39	45	49	42	V	0·08
Conisbrough	60	53	57	63	V	0·006
Oliver's Mount	60	53	57	63	V	1
Hunmanby	43	50	40	46	V	0·06
Belmont	25	32	22	28	H	500
Weaverthorpe	59	65	55	62	V	0·045

Conversion factors

To convert from column one to column two *multiply* by the conversion factor.

To convert	*Into*	*Multiply by*
acres	square feet	$4{\cdot}356 \times 10^{4}$
acres	square metres	4047
acres	square yards	$4{\cdot}84 \times 10^{3}$
acres	hectares	0·4047
ampere-hours	coulombs	3600
amperes per sq cm	amperes per sq inch	6·452
ampere-turns	gilberts	1·257
ampere-turns per cm	ampere-turns per inch	2·540
angstroms	nanometres	10^{-1}
ares	square metres	10^{2}
atmospheres	bars	1·0133
atmospheres	mm of mercury at 0°C	760
atmospheres	feet of water at 4°C	33·90
atmospheres	inches of mercury at 0°C	29·92
atmospheres	kg per sq metre	$1{\cdot}033 \times 10^{4}$
atmospheres	newtons per sq metre	$1{\cdot}0133 \times 10^{5}$
atmospheres	pounds per sq inch	14·70
barns	square metres	10^{-28}
bars	newtons per sq metre	10^{5}
bars	hectopiezes	1
bars	baryes (dyne per sq cm)	10^{6}
bars	pascals (newtons per sq metre)	10^{5}
baryes	newtons per sq metre	10^{-1}
Btu	foot-pounds	778·3
Btu	joules	1054·8
Btu	kilogram-calories	0·2520
Btu	horsepower-hours	$3{\cdot}929 \times 10^{-4}$
bushels	cubic feet	1·2445
calories (I.T.)	joules	4·1868
calories (thermochem)	joules	4·184
carats (metric)	grams	0·2
Celsius (centigrade)	Fahrenheit (see pages 158–159)	
chains (surveyor's)	feet	66
circular mils	square centimetres	$5{\cdot}067 \times 10^{-6}$
circular mils	square mils	0·7854
cords	cubic metres	3·625
cubic feet	cords	$7{\cdot}8125 \times 10^{-3}$
cubic feet	litres	28·32
cubic inches	cubic centimetres	16·39
cubic inches	cubic feet	$5{\cdot}787 \times 10^{-4}$
cubic inches	cubic metres	$1{\cdot}639 \times 10^{-5}$
cubic metres	cubic feet	35·31
cubic metres	cubic yards	1·308
degrees (angle)	radians	$1{\cdot}745 \times 10^{-2}$
dynes	pounds	$2{\cdot}248 \times 10^{-6}$
dynes	newtons	10^{-5}

To convert	*Into*	*Multiply by*
electron volts	joules	$1{\cdot}602 \times 10^{-19}$
ergs	foot-pounds	$7{\cdot}376 \times 10^{-8}$
ergs	joules	10^{-7}
fathoms	feet	6
fathoms	metres	1·8288
feet	centimetres	30·48
feet	varas	0·3594
feet of water at 4°C	inches of mercury at 0°C	0·8826
feet of water at 4°C	kg per sq metre	304·8
feet of water at 4°C	pounds per sq foot	62·43
fermis	metres	10^{-15}
footcandles	lumens per sq metre	10·764
footlamberts	candelas per sq metre	3·4263
foot-pounds	horsepower-hours	$5{\cdot}050 \times 10^{-7}$
foot-pounds	kilogram-metres	0·1383
foot-pounds	kilowatt-hours	$3{\cdot}766 \times 10^{-7}$
gallons (liq US)	gallons (liq Imp)	0·8327
gammas	teslas	10^{-9}
gausses	lines per sq inch	6·452
gausses	teslas	10^{-4}
gilberts	amperes	$7{\cdot}9577 \times 10^{-1}$
grain (for humidity calculations)	pounds (avoirdupois)	$1{\cdot}429 \times 10^{-4}$
grams	dynes	980·7
grams	grains	15·43
grams	ounces (avoirdupois)	$3{\cdot}527 \times 10^{-2}$
grams	poundals	$7{\cdot}093 \times 10^{-2}$
grams per cm	pounds per inch	$5{\cdot}600 \times 10^{-3}$
grams per cu cm	pounds per cu inch	$3{\cdot}613 \times 10^{-2}$
grams per sq cm	pounds per sq foot	2·0481
hectares	square metres	10^{4}
hectares	acres	2·471
horsepower (boiler)	Btu per hour	$3{\cdot}347 \times 10^{4}$
horsepower (metric) (542·5 ft-lb per second)	Btu per minute	41·83
horsepower (metric) (542·5 ft-lb per second)	foot-lb per minute	$3{\cdot}255 \times 10^{4}$
horsepower (metric) (542·5 ft-lb per second)	kg-calories per minute	10·54
horsepower (550 ft-lb per second)	Btu per minute	42·41
horsepower (550 ft-lb per second)	foot-lb per minute	$3{\cdot}3 \times 10^{4}$
horsepower (550 ft-lb per second)	kilowatts	0·745
horsepower (metric) (542·5 ft-lb per second)	horsepower (550 ft-lb per second)	0·9863
horsepower (550 ft-lb per second)	kg-calories per minute	10·69
inches	centimetres	2·540
inches	feet	$8{\cdot}333 \times 10^{-2}$
inches	miles	$1{\cdot}578 \times 10^{-5}$
inches	mils	1000
inches	yards	$2{\cdot}778 \times 10^{-2}$
inches of mercury at 0°C	lbs per sq inch	0·4912
inches of water at 4°C	kg per sq metre	25·40

To convert	*Into*	*Multiply by*
inches of water at 4°C	ounces per sq inch	0·5782
inches of water at 4°C	pounds per sq foot	5·202
inches of water at 4°C	in of mercury	$7{\cdot}355 \times 10^{-2}$
inches per ounce	metres per newton (compliance)	$9{\cdot}136 \times 10^{-2}$
joules	foot-pounds	0·7376
joules	ergs	10^{7}
kilogram-calories	kilogram-metres	426·9
kilogram-calories	kilojoules	4·186
kilogram-metres	joules	0·102
kilogram force	newtons	0·102
kilograms	tons, long (avdp 2240 lb)	$9{\cdot}842 \times 10^{-4}$
kilograms	tons, short (avdp 2000 lb)	$1{\cdot}102 \times 10^{-3}$
kilograms	pounds (avoirdupois)	2·205
kilograms per kilometre	pounds (avdp) per mile (stat)	3·548
kg per sq metre	pounds per sq foot	0·2048
kilometres	feet	3281
kilopond force	newtons	9·81
kilowatt-hours	Btu	3413
kilowatt-hours	foot-pounds	$2{\cdot}655 \times 10^{6}$
kilowatt-hours	joules	$3{\cdot}6 \times 10^{6}$
kilowatt-hours	kilogram-calories	860
kilowatt-hours	kilogram-metres	$3{\cdot}671 \times 10^{5}$
kilowatt-hours	pounds carbon oxidized	0·235
kilowatt-hours	pounds water evaporated from and at 212°F	3·53
kilowatt-hours	pounds water raised from 62° to 212°F	22·75
kips	newtons	$4{\cdot}448 \times 10^{3}$
knots* (naut mi per hour)	feet per second	1·688
knots	metres per minute	30·87
knots	miles (stat) per hour	1·1508
lamberts	candelas per sq cm	0·3183
lamberts	candelas per sq inch	2·054
lamberts	candelas per sq metre	$3{\cdot}183 \times 10^{3}$
leagues	miles (approximately)	3
links (surveyor's)	chains	0·01
links	inches	7·92
litres	bushels (dry US)	$2{\cdot}838 \times 10^{-2}$
litres	cubic centimetres	1000
litres	cubic metres	0·001
litres	cubic inches	61·02
litres	gallons (liq Imp)	0·2642
litres	pints (liq Imp)	1·816
$\log_e$ or ln	$\log_{10}$	0·4343
lumens per sq foot	foot-candles	1
lux	lumens per sq foot	0·0929
maxwells	webers	10^{-8}
metres	yards	1·094
metres	varas	1·179
metres per min	feet per minute	3·281
metres per min	kilometres per hour	0·06
microhms per cu cm	microhms per inch cube	0·3937
microhms per cu cm	ohms per mil foot	6·015
microns	metres	10^{-6}
miles (nautical)*	feet	6076·1

To convert	*Into*	*Multiply by*
miles (nautical)	metres	1852
miles (nautical)	miles (statute)	1·1508
miles (statute)	feet	5280
miles (statute)	kilometres	1·609
miles per hour	kilometres per minute	$2{\cdot}682 \times 10^{-2}$
miles per hour	feet per minute	88
miles per hour	kilometres per hour	1·609
millibars	inches of mercury (0°C)	0·02953
millibars (10^3 dynes per sq cm)	pounds per sq foot	2·089
mils	metres	$2{\cdot}54 \times 10^{-5}$
nepers	decibels	8·686
newtons	dynes	10^5
newtons	kilograms	0·1020
newtons	poundals	7·233
newtons	pounds (avoirdupois)	0·2248
oersteds	amperes per metre	$7{\cdot}9577 \times 10$
ounce-inches	newton-metres	$7{\cdot}062 \times 10^{-3}$
ounces (fluid)	quarts	$3{\cdot}125 \times 10^{-2}$
ounces (avoirdupois)	pounds	$6{\cdot}25 \times 10^{-2}$
pascals	newtons per sq metre	1
pascals	pounds per sq inch	$1{\cdot}45 \times 10^{-4}$
piezes	newtons per sq metre	10^3
piezes	sthenes per sq metre	1
pints	quarts	0·50
poises	newton-seconds per sq metre	10^{-1}
pounds of water (dist)	cubic feet	$1{\cdot}603 \times 10^{-2}$
pounds per inch	kg per metre	17·86
pounds per foot	kg per metre	1·488
pounds per mile (statute)	kg per kilometre	0·2818
pounds per cu foot	kg per cu metre	16·02
pounds per cu inch	pounds per cu foot	1728
pounds per sq foot	pounds per sq inch	$6{\cdot}944 \times 10^{-3}$
pounds per sq foot	kg per sq metre	4·882
pounds per sq inch	kg per sq metre	703·1
poundals	dynes	$1{\cdot}383 \times 10^4$
poundals	pounds (avoirdupois)	$3{\cdot}108 \times 10^{-2}$
quarts	gallons	0·25
rods	feet	16·5
slugs (mass)	pounds (avoirdupois)	32·174
sq inches	circular mils	$1{\cdot}273 \times 10^6$
sq inches	sq centimetres	6·452
sq feet	sq metres	$9{\cdot}290 \times 10^{-2}$
sq miles	sq yards	$3{\cdot}098 \times 10^6$
sq miles	acres	640
sq miles	sq kilometres	2·590
sq millimetres	circular mils	1973
steres	cubic metres	1
stokes	sq metres per second	10^{-4}
(temp rise, °C) × (US gal water)/minute	watts	264
tonnes	kilograms	10^3
tons, short (avoir 2000 lb)	tonnes (1000 kg)	0·9072
tons, long (avoir 2240 lb)	tonnes (1000 kg)	1·016
tons, long (avoir 2240 lb)	tons, short (avoir 2000 lb)	1·120
tons (US shipping)	cubic feet	40

To convert	*Into*	*Multiply by*
torrs	newtons per sq metre	133·32
watts	Btu per minute	$5{\cdot}689 \times 10^{-2}$
watts	ergs per second	10^{7}
watts	foot-lb per minute	44·26
watts	horsepower (550 ft-lb per second)	$1{\cdot}341 \times 10^{-3}$
watts	horsepower (metric) (542·5 ft-lb per second)	$1{\cdot}360 \times 10^{-3}$
watts	kg-calories per minute	$1{\cdot}433 \times 10^{-2}$
watt-seconds (joules)	gram-calories (mean)	0·2389
webers per sq metre	gausses	10^{4}
yards	feet	3

Fractions of an inch with metric equivalents

Fractions of an inch		Decimals of an inch	mm
	1/64	0·0156	0·397
1/32		0·0312	0·794
	3/64	0·0468	1·191
1/16		0·0625	1·588
	5/64	0·0781	1·985
3/32		0·0938	2·381
	7/64	0·1094	2·778
1/8		0·1250	3·175
	9/64	0·1406	3·572
5/32		0·1563	3·969
	11/64	0·1719	4·366
3/16		0·1875	4·762
	13/64	0·2031	5·159
7/32		0·2187	5·556
	15/64	0·2344	5·953
1/4		0·2500	6·350
	17/64	0·2656	6·747
9/32		0·2813	7·144
	19/64	0·2969	7·541
5/16		0·3125	7·937
	21/64	0·3281	8·334
11/32		0·3438	8·731
	23/64	0·3593	9·128
3/8		0·3750	9·525
	25/64	0·3906	9·922
13/32		0·4063	10·319
	27/64	0·4219	10·716
7/16		0·4375	11·112
	29/64	0·4531	11·509
15/32		0·4687	11·906
	31/64	0·4844	12·303
1/2		0·5000	12·700
	33/64	0·5156	13·097
17/32		0·5313	13·494
	35/64	0·5469	13·891
9/16		0·5625	14·287
	37/64	0·5781	14·684
19/32		0·5938	15·081
	39/64	0·6094	15·478
5/8		0·6250	15·875
	41/64	0·6406	16·272
21/32		0·6563	16·668
	43/64	0·6719	17·065
11/16		0·6875	17·462
	45/64	0·7031	17·859
23/32		0·7188	18·256
	47/64	0·7344	18·653
3/4		0·7500	19·050
	49/64	0·7656	19·447
25/32		0·7813	19·843
	51/64	0·7969	20·240
13/16		0·8125	20·637
	53/64	0·8281	21·034
27/32		0·8438	21·431
	55/64	0·8594	21·828
7/8		0·8750	22·225
	57/64	0·8906	22·622
29/32		0·9062	23·019
	59/64	0·9219	23·416
15/16		0·9375	23·812

31/32		0·9688	24·606
	61/64	0·9531	24·209
	63/64	0·9844	25·003
		1·000	25·400

Code conversion tables

Dec	Octal	Hex	Binary bit pattern 7 6 5 4 3 2 1	ASCII character
0	0	0	0 0 0 0 0 0 0	
1	1	1	0 0 0 0 0 0 1	
2	2	2	0 0 0 0 0 1 0	
3	3	3	0 0 0 0 0 1 1	
4	4	4	0 0 0 0 1 0 0	
5	5	5	0 0 0 0 1 0 1	
6	6	6	0 0 0 0 1 1 0	
7	7	7	0 0 0 0 1 1 1	
8	10	8	0 0 0 1 0 0 0	
9	11	9	0 0 0 1 0 0 1	
10	12	A	0 0 0 1 0 1 0	
11	13	B	0 0 0 1 0 1 1	
12	14	C	0 0 0 1 1 0 0	
13	15	D	0 0 0 1 1 0 1	
14	16	E	0 0 0 1 1 1 0	Special characters (0–31)
15	17	F	0 0 0 1 1 1 1	
16	20	10	0 0 1 0 0 0 0	
17	21	11	0 0 1 0 0 0 1	
18	22	12	0 0 1 0 0 1 0	
19	23	13	0 0 1 0 0 1 1	
20	24	14	0 0 1 0 1 0 0	
21	25	15	0 0 1 0 1 0 1	
22	26	16	0 0 1 0 1 1 0	
23	27	17	0 0 1 0 1 1 1	
24	30	18	0 0 1 1 0 0 0	
25	31	19	0 0 1 1 0 0 1	
26	32	1A	0 0 1 1 0 1 0	
27	33	1B	0 0 1 1 0 1 1	
28	34	1C	0 0 1 1 1 0 0	
29	35	1D	0 0 1 1 1 0 1	
30	36	1E	0 0 1 1 1 1 0	
31	37	1F	0 0 1 1 1 1 1	
32	40	20	0 1 0 0 0 0 0	SPACE
33	41	21	0 1 0 0 0 0 1	!
34	42	22	0 1 0 0 0 1 0	"
35	43	23	0 1 0 0 0 1 1	#
36	44	24	0 1 0 0 1 0 0	$
37	45	25	0 1 0 0 1 0 1	%
38	46	26	0 1 0 0 1 1 0	&
39	47	27	0 1 0 0 1 1 1	'
40	50	28	0 1 0 1 0 0 0	(
41	51	29	0 1 0 1 0 0 1	)
42	52	2A	0 1 0 1 0 1 0	*
43	53	2B	0 1 0 1 0 1 1	+

Dec	Octal	Hex	Binary bit pattern	ASCII character
44	54	2C	0 1 0 1 1 0 0	,
45	55	2D	0 1 0 1 1 0 1	—
46	56	2E	0 1 0 1 1 1 0	.
47	57	2F	0 1 0 1 1 1 1	/
48	60	30	0 1 1 0 0 0 0	Ø
49	61	31	0 1 1 0 0 0 1	1
50	62	32	0 1 1 0 0 1 0	2
51	63	33	0 1 1 0 0 1 1	3
52	64	34	0 1 1 0 1 0 0	4
53	65	35	0 1 1 0 1 0 1	5
54	66	36	0 1 1 0 1 1 0	6
55	67	37	0 1 1 0 1 1 1	7
56	70	38	0 1 1 1 0 0 0	8
57	71	39	0 1 1 1 0 0 1	9
58	72	3A	0 1 1 1 0 1 0	:
59	73	3B	0 1 1 1 0 1 1	;
60	74	3C	0 1 1 1 1 0 0	<
61	75	3D	0 1 1 1 1 0 1	=
62	76	3E	0 1 1 1 1 1 0	>
63	77	3F	0 1 1 1 1 1 1	?
64	100	40	1 0 0 0 0 0 0	@
65	101	41	1 0 0 0 0 0 1	A
66	102	42	1 0 0 0 0 1 0	B
67	103	43	1 0 0 0 0 1 1	C
68	104	44	1 0 0 0 1 0 0	D
69	105	45	1 0 0 0 1 0 1	E
70	106	46	1 0 0 0 1 1 0	F
71	107	47	1 0 0 0 1 1 1	G
72	110	48	1 0 0 1 0 0 0	H
73	111	49	1 0 0 1 0 0 1	I
74	112	4A	1 0 0 1 0 1 0	J
75	113	4B	1 0 0 1 0 1 1	K
76	114	4C	1 0 0 1 1 0 0	L
77	115	4D	1 0 0 1 1 0 1	M
78	116	4E	1 0 0 1 1 1 0	N
79	117	4F	1 0 0 1 1 1 1	O
80	120	50	1 0 1 0 0 0 0	P
81	121	51	1 0 1 0 0 0 1	Q
82	122	52	1 0 1 0 0 1 0	R
83	123	53	1 0 1 0 0 1 1	S
84	124	54	1 0 1 0 1 0 0	T
85	125	55	1 0 1 0 1 0 1	U
86	126	56	1 0 1 0 1 1 0	V
87	127	57	1 0 1 0 1 1 1	W
88	130	58	1 0 1 1 0 0 0	X
89	131	59	1 0 1 1 0 0 1	Y
90	132	5A	1 0 1 1 0 1 0	Z
91	133	5B	1 0 1 1 0 1 1	[
92	134	5C	1 0 1 1 1 0 0	\
93	135	5D	1 0 1 1 1 0 1	]
94	136	5E	1 0 1 1 1 1 0	↑
95	137	5F	1 0 1 1 1 1 1	←
96	140	60	1 1 0 0 0 0 0	—
97	141	61	1 1 0 0 0 0 1	a

Dec	Octal	Hex	Binary bit pattern	ASCII character
98	142	62	1 1 0 0 0 1 0	b
99	143	63	1 1 0 0 0 1 1	c
100	144	64	1 1 0 0 1 0 0	d
101	145	65	1 1 0 0 1 0 1	e
102	146	66	1 1 0 0 1 1 0	f
103	147	67	1 1 0 0 1 1 1	g
104	150	68	1 1 0 1 0 0 0	h
105	151	69	1 1 0 1 0 0 1	i
106	152	6A	1 1 0 1 0 1 0	j
107	153	6B	1 1 0 1 0 1 1	k
108	154	6C	1 1 0 1 1 0 0	l
109	155	6D	1 1 0 1 1 0 1	m
110	156	6E	1 1 0 1 1 1 0	n
111	157	6F	1 1 0 1 1 1 1	o
112	160	70	1 1 1 0 0 0 0	p
113	161	71	1 1 1 0 0 0 1	q
114	162	72	1 1 1 0 0 1 0	r
115	163	73	1 1 1 0 0 1 1	s
116	164	74	1 1 1 0 1 0 0	t
117	165	75	1 1 1 0 1 0 1	u
118	166	76	1 1 1 0 1 1 0	v
119	167	77	1 1 1 0 1 1 1	w
120	170	78	1 1 1 1 0 0 0	x
121	171	79	1 1 1 1 0 0 1	y
122	172	7A	1 1 1 1 0 1 0	z
123	173	7B	1 1 1 1 0 1 1	—
124	174	7C	1 1 1 1 1 0 0	—
125	175	7D	1 1 1 1 1 0 1	—
126	176	7E	1 1 1 1 1 1 0	—
127	177	7F	1 1 1 1 1 1 1	—

— means special characters or codes not used

Musical notes frequency

The range of notes on a piano keyboard is from 27·5 Hz to 4186 Hz. Middle C (the centre note on a standard keyboard) has a frequency of 261·6 Hz. Standard pitch is A above middle C at a frequency of 440 Hz. Note that raising the pitch of a note is equivalent to doubling the frequency for each complete octave.

A	27·5	D	73·4	G	196·0	C	523·3	F	1396·9	B	3951·1
B	30·9	E	82·4	A	220·0	D	587·3	G	1568·0	C	4186·0
C	32·7	F	87·3	B	246·9	E	659·2	A	1760·0		
D	36·7	G	98·0	C	261·6	F	698·5	B	1975·5		
E	41·2	A	110·0	D	293·7	G	784·0	C	2093·0		
F	43·7	B	123·5	E	329·6	A	880·0	D	2344·3		
G	49·0	C	130·8	F	349·2	B	987·8	E	2637·0		
A	55·0	D	146·8	G	392·0	C	1046·5	F	2793·8		
B	61·7	E	164·8	A	440·0	D	1174·0	G	3136·0		
C	65·4	F	174·6	B	493·9	E	1318·5	A	3520·0		

Semiconductor glossary

The more common terms relating to semiconductors are explained briefly here. Space naturally prevents a complete explanation and only allows for inclusion of the most common terms.

Admittance Reciprocal of impedance, symbol Y. The unit of admittance is the *Siemen*.

Alpha (α) The term used for the current gain of a transistor in the common-base mode. A term now rarely used.

Beta (β) The current gain of a transistor in the common-emitter mode. Since the introduction of the hybrid parameters system the terms h_{fe} (the small signal current gain) or h_{FE} (the d.c. gain, which is the collector current divided by the base current) are generally used for gain in the common-emitter mode.

Bias For a transistor to operate correctly the proper potentials have to be present at its emitter, base and collector. Normally the term bias refers to the voltage applied to the base to bring the operating point to a linear part of the amplification curve. For germanium transistors this is usually 0·3 V with respect to the emitter and for silicon transistors at least 0·6 V.

Complementary pair Most modern transistor audio amplifiers make use of a pair of transistors, one npn and the other pnp, with similar characteristics and closely matched gains in the driver or output stage: they are referred to as a complementary pair.

Darlington pair Circuit using two transistors with the collectors connected together and the emitter of the first directly coupled to the base of the second. This configuration gives very high gains—equal to the gains of the two individual transistors multiplied together.

Diac Bi-directional voltage breakdown diode; passes current above a certain breakdown voltage. Normally employed with a triac in an a.c. control circuit.

Diode (semiconductor) Simple pn junction device which presents a high resistance one way around and a low resistance the other. Well known as a detector but with a wide variety of applications in all fields of electronics.

F.E.T. (field effect transistor) The f.e.t. makes use of the electric field established in a p- or n-type channel of semiconductor material to control the flow of current through the channel. The field is established by the bias applied to the gate connections and the f.e.t. is thus a voltage-controlled device. This means that it has a much higher input impedance than ordinary transistors. The main connections are the source, drain and gate but some f.e.t.s have additional connections.

Forward bias The biasing of a pn junction so that conduction increases. This occurs when positive connects to the p-side with negative to the n-side of the junction.

Germanium One of the two main semiconductor elements. For it to exhibit semiconductor characteristics, it must first be purified and then tri- or pentavalent impurities in minute but carefully controlled quantities added to give p- or n-type semiconductor material respectively.

Hall effect device A device which, when current is passed through it, develops an e.m.f. when placed in a magnetic field.

Heat sink Current flow through a semiconductor device results in the production of heat. In low power devices no precautions are needed in normal operation as the heat can be dissipated by the body of the device. High power devices, however, require help in the form of a heat conductor to get rid of the excess heat. This may take the form of a fin clamped around the body but in others the body is designed to be bolted to a metal plate. Since semiconductor devices are heat sensitive the current through them increases with increase in heat. It is also necessary to take precautions when soldering a semiconductor into circuit.

h_{fe} and h_{FE} See Beta.

Holding current Lowest current at which a thyristor-type device will continue to pass current after gate voltage has been removed.

Hybrid circuit Circuit employing both transistors and valves or transistors and integrated circuits.

Integrated circuit An integrated circuit consists of transistors, diodes, resistors and capacitors with the necessary interconnections laid out to form a circuit such as an amplifier, bistable switch, etc. and fabricated on a single chip of semiconductor material. The chips themselves are extremely small—a fraction of the size of the encapsulation.

Leadout wires from the i.c. itself to the package pins are attached for connection of the inputs, outputs, supply voltages and any components that cannot be included such as controls, high value capacitors and inductors. These additional components are termed 'discrete components'.

Junction capacitance Capacitance between pn junctions in a semiconductor device. Also called barrier, depletion layer or transition capacitances (see Neutralisation).

Light sensitive devices Light and heat both affect the conductivity of a pn junction. Devices are available in which a pn junction is exposed to light so as to make use of this property. Light falling on the junction liberates current carriers and allows the device to conduct.

M.O.S.T. Type of f.e.t. with oxide insulating layer between the metal gate and semiconductor channel. It has a higher input impedance than the junction type f.e.t.

N-type semiconductor material Silicon or germanium doped with pentavalent impurity to give an excess of negative current carriers (i.e. free electrons).

Neutralisation In radio frequency transistors there is a tendency for self-oscillation to occur due to the collector-base capacitance. In modern r.f. transistors this capacitance can be made very small. To overcome the effect in early r.f. transistors it was usual to use a small amount of capacitive negative feedback in each stage, this being known as neutralisation.

PN junction Junction between p- and n-type semiconductor material with the semiconductor crystal structure.

Point contact device One in which the pn junction is formed at the contact between a metal 'cats-whisker' and the semiconductor material. Point contact diodes have advantages in some applications.

P-type semiconductor material Semiconductor material doped with trivalent impurity to give an excess of positive current carriers (i.e. holes).

Ratings Specification sheets for transistors cover many facets of the device's operation but most parameters are needed only by the designer. The ratings which need to be known for replacement purposes are $V_{CE(max)}$, the maximum collector to emitter voltage; I_C the collector current; h_{fe}, the gain and f_t the cut-off frequency. The output power also needs to be observed.

Reverse bias Bias applied to a pn junction so as to reduce the current flow through it—positive connects to the n-type and negative to the p-type semiconductor material.

Saturation Of a transistor when the collector current is limited by the external circuits and not by the base bias applied.

Semiconductor device Device whose operation is based on the use of semiconductor material. In addition to transistors and diodes there are a wide range of components which make use of semiconductor effects.

Semiconductor material Material whose conductive properties depend on the addition of minute quantities of impurity atoms. Pure germanium has a conductivity of about 60 ohms per cm^3 and silicon about 60,000 ohms per cm^3; semiconductors need to be about 2 ohms per cm^3. Addition of tri- or pentavalent impurities produce p- or n-type semiconductor material with an excess of positive or negative current carriers. Unlike normal conductors, semiconductors increase in conductivity with an increase in temperature.

Silicon One of the two main semiconductor materials in use. Its energy gap (1·08eV) makes it less sensitive to heat than germanium.

Solid state circuit A circuit in which the current flows through solid material instead of through a gas or vacuum.

Super alpha pair See Darlington pair.

Thermal runaway Semiconductor materials are very sensitive to heat—germanium much more so than silicon. Circuit design has to take account of this and many components have to be included to prevent increased current flow due to heat. Without such protection heat induced current will raise the temperature leading to a further increase in current and so on, a process known as thermal runaway which can destroy a semiconductor. See Heat sink.

Thermistor A semiconductor whose resistance varies with temperature. Some thermistors have a negative temperature coefficient, that is resistance falls with an increase in temperature, others have a positive temperature coefficient. Used to compensate for the effect of heat in circuit operation.

Thyristor A three-junction, four-layer semiconductor rectifier which conducts when either the voltage across it reaches a 'breakdown' point or when triggered by a pulse fed to its gate electrode. Once triggered by a pulse it remains conducting until the a.c. voltage across it reverses phase.

Triac Bi-directional thyristor. Used in a.c. control circuitry.

Tunnel diode A heavily doped semiconductor diode which exhibits a negative-resistance characteristic, i.e. over part of its characteristic increased forward bias leads to a reduction in the current flowing.

Type numbers The numbers in a transistor designation rarely describe anything about its characteristics. In the 2N series adjacent type numbers are frequently widely differing devices. European and British transistors are frequently coded with the first letter A (germanium) or B (silicon) followed by a second letter which indicates the type:

A	Diode	P	Photo type
C	a.f. (low power)	S	Switching (low power)
D	a.f. (power)	U	Switching (power)
E	Tunnel diode	Y	Diode (power)
F	h.f. (low power)	Z	Zener diode
L	h.f. (power)		

Unijunction transistor Three-terminal transistor consisting of an n-type silicon bar with a base contact at each end called 'base 1' and 'base 2' and a p-type emitter region with further contact at one side. Current flow through the device from one base to the other is controlled by the current fed to the emitter. When the emitter voltage reaches a certain level, the emitter-base 1 junction virtually short-circuits. With a suitable charging circuit at the emitter the device operates as a relaxation oscillator.

Valency The ability of atoms to unite with other atoms; due to the electrons that exist in the outer orbit, or valency band, being able to form a shared orbit with other atoms.

Varicap diode When reverse biased, all pn junctions exhibit capacitance, since the depletion layer at the junction forms an insulator between the conductive regions. This property is exploited by the varicap diode which is used for such purposes as automatic tuning and a.f.c.

Voltage-dependent resistor Resistor using semiconductor material whose resistance varies with applied voltage.

Zener diode Junction diode designed to operate reverse biased into the breakdown region of its characteristic. In this region large increases in current produce negligible variation in the voltage across it. Therefore it can be used for voltage stabilisation or to establish a stable reference voltage.

Using LEDs

When using a light-emitting diode (LED) as an indicator, use the following formula to determine series resistance for various voltages: $R = (V - 1{\cdot}7) \times 1000 \div I$, where R is resistance in ohms, V is supply voltage (d.c.!), and I is LED current in milliamps.

E.g. to operate LED at 20 mA on

6 V	use 220 ohms
9 V	390
12 V	560
24 V	1·2k

To operate a LED directly from the 240 V mains, a better scheme is to use the second circuit shown. In this, a capacitor is used as a voltage dropping element. A 1N4148 diode or similar across the LED provides the rectification required. As the voltage drop across the LED is negligible compared with the supply, capacitor current is almost always exactly equal to mains voltage divided by capacitive reactance X_c.

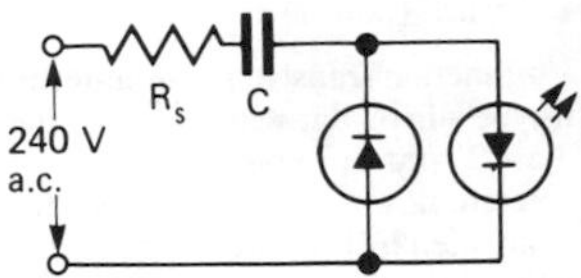

At 50 Hz, 0·47 µF will result in a LED current of about 16 mA. Resistor R_s is included to limit turn-on transients. A value of 270 ohms should be adequate.

Power supply configurations

No circuit losses are allowed for. At low voltages allow for 0·6 V diode drop.

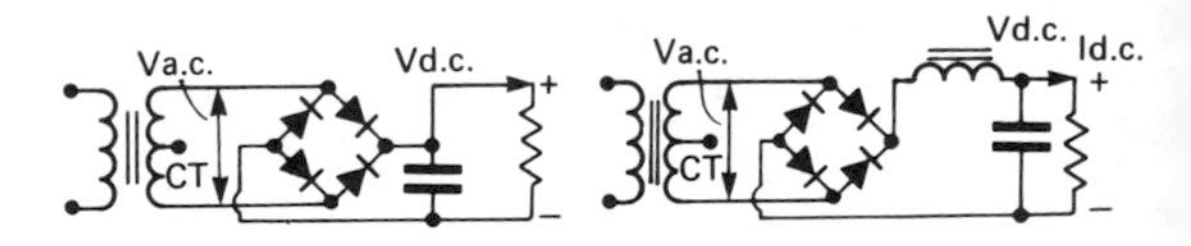

Full-wave bridge
Capacitive input filter

$V_{d.c.} = 1{\cdot}41 \times V_{a.c.}$

$I_{d.c.} = 0{\cdot}62 \times I_{a.c.}$

Full-wave bridge
Choke input filter

$V_{d.c.} = 0{\cdot}90 \times V_{a.c.}$

$I_{d.c.} = 0{\cdot}94 \times I_{a.c.}$

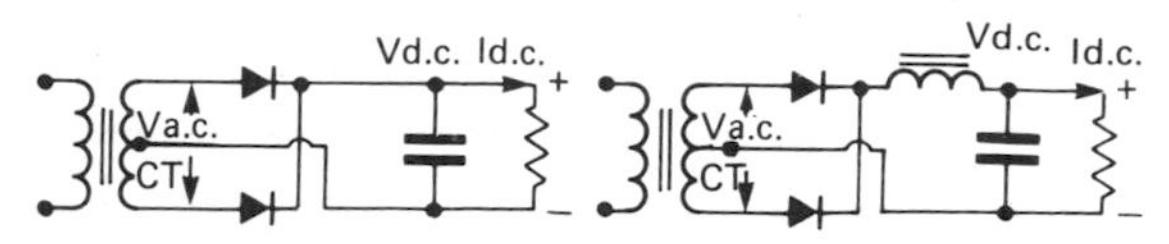

Full-wave
Capacitive input filter

$V_{d.c.} = 0{\cdot}71 \times V_{a.c.}$

$I_{d.c.} = 1{\cdot}0 \times I_{a.c.}$

Full-wave
Choke input filter

$V_{d.c.} = 0{\cdot}45 \times V_{a.c.}$

$I_{d.c.} = 1{\cdot}54 \times I_{a.c.}$

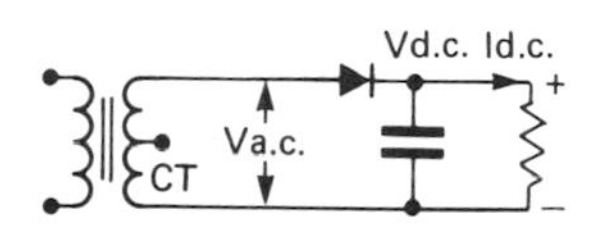

Half-wave
Capacitive input filter

$V_{d.c.} = 1{\cdot}41 \times V_{a.c.}$

$I_{d.c.} = 0{\cdot}28 \times I_{a.c.}$

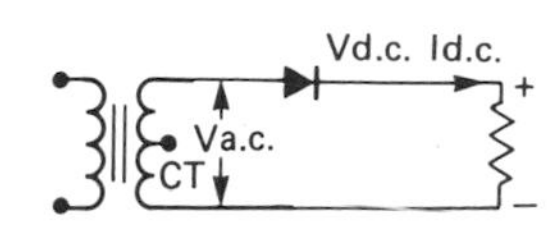

Half-wave
Resistive load

$V_{d.c.} = 0{\cdot}45 \times V_{a.c.}$

$I_{d.c.} = 0{\cdot}64 \times I_{a.c.}$

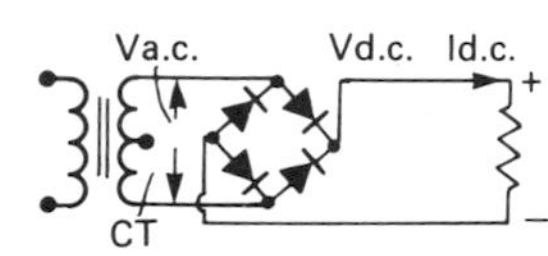

Full-wave bridge
Resistive load

$V_{d.c.} = 0{\cdot}90 \times V_{a.c.}$

$I_{d.c.} = 0{\cdot}90 \times I_{a.c.}$

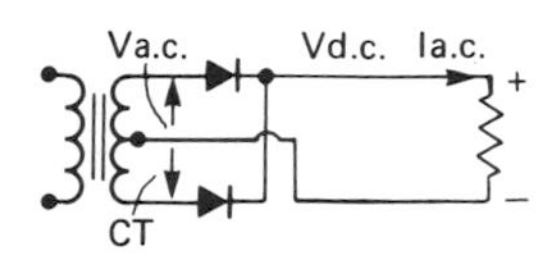

Full-wave
Resistive load

$V_{d.c.} = 0{\cdot}45 \times V_{a.c.}$

$I_{d.c.} = 1{\cdot}27 \times I_{a.c.}$

Voltage multiplier circuits

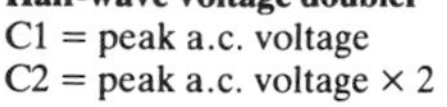

Half-wave voltage doubler
C1 = peak a.c. voltage
C2 = peak a.c. voltage × 2

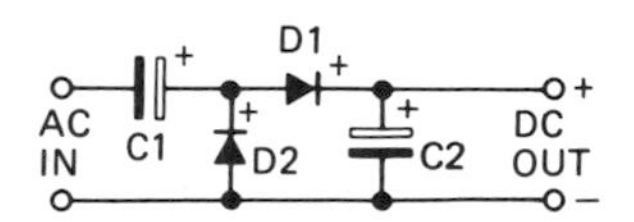

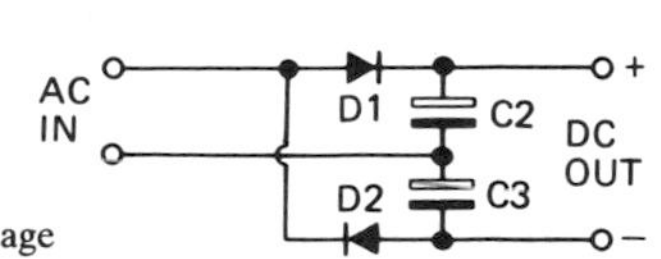

Bi-phase half wave or full wave voltage doubler
C2 and C3 = peak a.c. voltage

Voltage tripler
C1 = peak a.c. voltage
C2 = peak a.c. voltage
C3 = peak a.c. voltage × 2

Voltage quadrupler
C1A = peak a.c. voltage
C1B = peak a.c. voltage × 3
C2A and C2B = peak a.c. voltage × 2
D1-D4 = peak a.c. voltage × 2

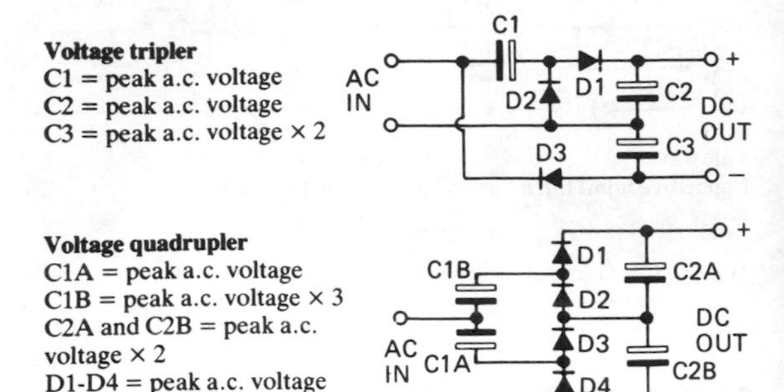

Zener diodes

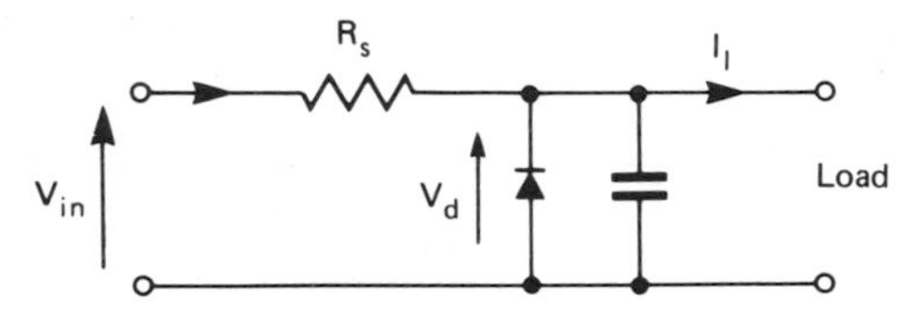

Constant load current/variable input voltage

$$\text{Series resistor } R_s = \frac{V_{in}(\min) - V_d}{1{\cdot}1 I_L}$$

$$\text{Diode dissipation } P_d = \left(\frac{V_{in}(\max) - V_d}{R_s} - I_L \right) V_d$$

Variable load current/constant input voltage

$$R_s = \frac{V_{in} - V_d}{1{\cdot}1 I_L(\max)}$$

$$P_d = \left(\frac{V_{in} - V_d}{R_s} - I_L(\min) \right) V_d$$

Variable load current/variable input voltage

$$R_s = \frac{V_{in}(\min) - V_d}{1{\cdot}1 I_L(\max)}$$

$$P_d = \left(\frac{V_{in}(\max) - V_d}{R_s} - I_L(\min) \right) V_d$$

Voltage regulators

Fixed voltage types eg 78, 79 series

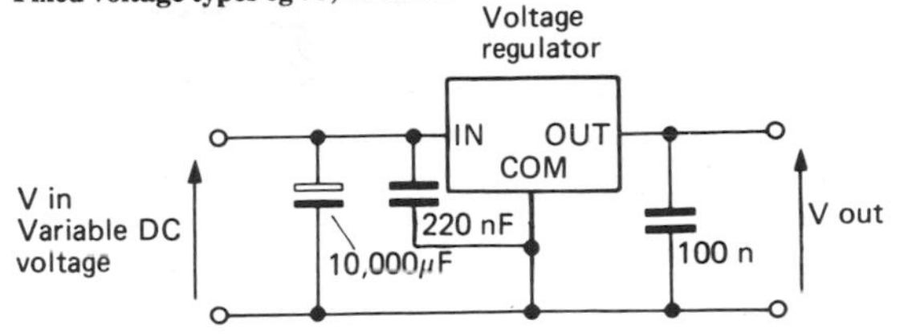

Variable voltage types eg 317, 338 series

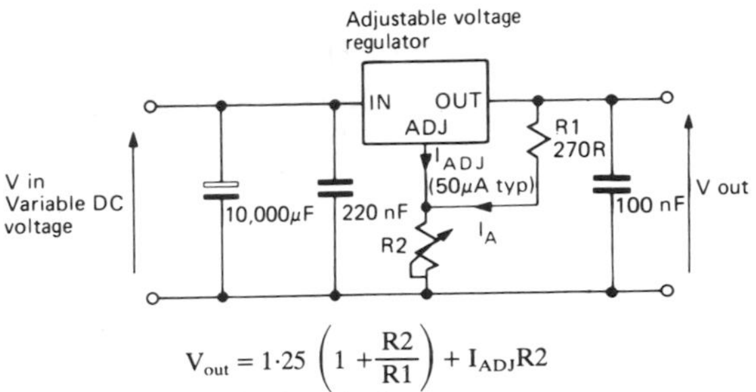

$$V_{out} = 1{\cdot}25\left(1 + \frac{R2}{R1}\right) + I_{ADJ}R2$$

Selcct R1 and R2 so that $I_A > 4\,mA$

Op-amp standard circuits

A_v = closed loop a.c. gain
e_i = input voltage
e_o = output voltage
f_o = low frequency -3 dB point
R_{in} = input impedance

Split supply configurations–supply connections omitted for clarity

Non-inverting a.c. amplifier

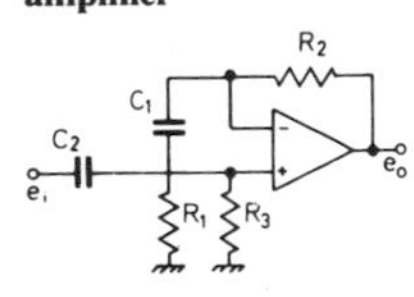

$$A_v = 1 + \frac{R_2}{R_1};\ R_{in} = R_2$$

$$f_o = \frac{1}{2\pi R_1 C_1} = \frac{1}{2\pi R_3 C_2}$$

Inverting a.c. amplifier

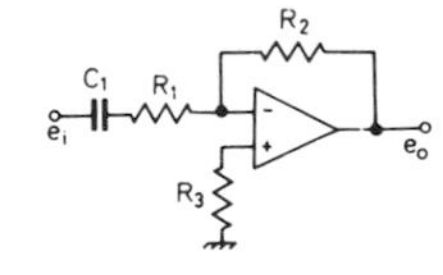

$$A_v = \frac{R_2}{R_1};\ R_{in} = R_1$$

$$f_o = \frac{1}{2\pi R_1 C_1}$$

Non-inverting buffer

$A_v = 1 \qquad R_{in} = R_1$

$$f_o = \frac{1}{2\pi R_1 C_1}$$

Inverting buffer

$A_v = -1 \qquad R_{in} = R_1$

$$f_o = \frac{1}{2\pi R_1 C_1}$$

Inverting summing amplifier

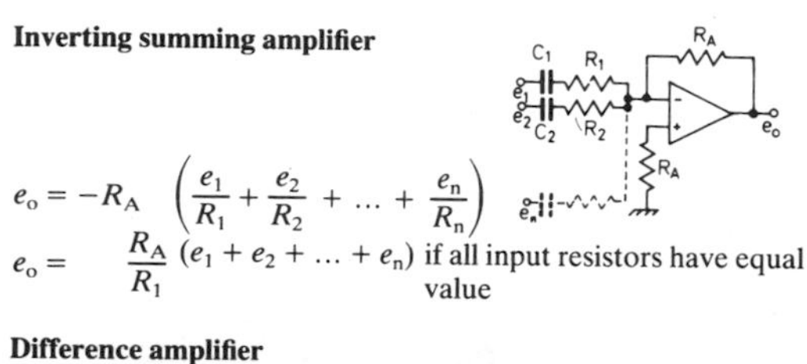

$$e_o = -R_A \left(\frac{e_1}{R_1} + \frac{e_2}{R_2} + \ldots + \frac{e_n}{R_n}\right)$$

$$e_o = \frac{R_A}{R_1}(e_1 + e_2 + \ldots + e_n)$$ if all input resistors have equal value

Difference amplifier

$$e_o = \left(\frac{R_1 + R_2}{R_3 + R_4}\right)\frac{R_4}{R_1} e_2 - \frac{R_2}{R_1} e_1$$

if $R_1 = R_3$ and $R_2 = R_4$ then

$$e_o = \frac{R_2}{R_1}(e_2 - e_1)$$

$$f_o = \frac{1}{2\pi R_1 C_1} = \frac{1}{2\pi(R_3 + R_4)C_3}$$

$R_2 = R_4$ for minimum offset error

Variable gain a.c. amplifier

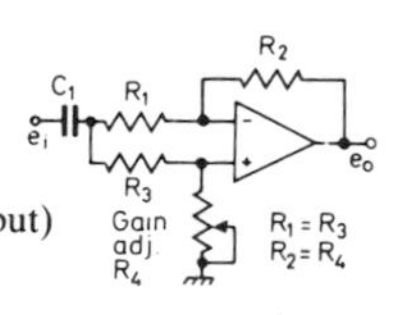

$A_v = 0$ (slider at ground)

$A_{v,max} = -\frac{R_2}{R_1}$ (slider at positive input)

$R_{in} = \frac{R_1}{2}$ (minimum)

$$f_o = \frac{1}{2\pi(\tfrac{1}{2}R_1)C_1}$$

Single supply configurations–supply connections omitted for clarity.

Polarity switcher, or 4-quadrant gain control

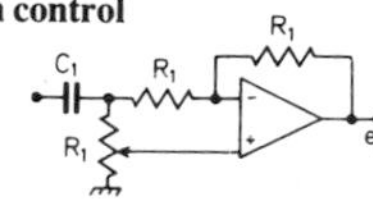

$A_v = +1$ (slider at C_1)

$A_v = 0$ (slider midposition)

$A_v = -1$ (slider at ground)

$R_{in} = \frac{R_1}{2}$ (minimum) $\qquad f_o = \frac{1}{2\pi(\tfrac{1}{2}R_1)C_1}$

Single supply biasing of non-inverting a.c. amplifier

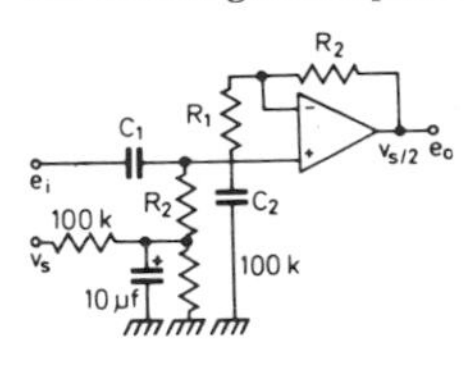

$$A_v = 1 + \frac{R_2}{R_1}$$

$$R_{in} = R_2$$

$$f_o = \frac{1}{2\pi R_2 C_1} = \frac{1}{2\pi R_1 C_2}$$

Single supply biasing of inverting a.c. amplifier

$$A_v = -\frac{R_2}{R_1} ; R_{in} = R_1$$

$$f_o = \frac{1}{2\pi R_1 C_1}$$

Decibel table

The decibel figures are in the centre column: figures to the left represent decibel loss, and those to the right decibel gain. The voltage and current figures are given on the assumption that there is no change in impedance.

Voltage or current ratio	*Power ratio*	*dB* ←− +→	*Voltage or current ratio*	*Power ratio*
1·000	1·000	0	1·000	1·000
0·989	0·977	0·1	1·012	1·023
0·977	0·955	0·2	1·023	1·047
0·966	0·933	0·3	1·035	1·072
0·955	0·912	0·4	1·047	1·096
0·944	0·891	0·5	1·059	1·122
0·933	0·871	0·6	1·072	1·148
0·912	0·832	0·8	1·096	1·202
0·891	0·794	1·0	1·122	1·259
0·841	0·708	1·5	1·189	1·413
0·794	0·631	2·0	1·259	1·585
0·750	0·562	2·5	1·334	1·778
0·708	0·501	3·0	1·413	1·995
0·668	0·447	3·5	1·496	2·239
0·631	0·398	4·0	1·585	2·512
0·596	0·355	4·5	1·679	2·818
0·562	0·316	5·0	1·778	3·162
0·501	0·251	6·0	1·995	3·981
0·447	0·200	7·0	2·239	5·012
0·398	0·159	8·0	2·512	6·310
0·355	0·126	9·0	2·818	7·943
0·316	0·100	10	3·162	10·00
0·282	0·0794	11	3·55	12·6
0·251	0·0631	12	3·98	15·9
0·224	0·0501	13	4·47	20·0
0·200	0·0398	14	5·01	25·1
0·178	0·0316	15	5·62	31·6

Voltage or current ratio	Power ratio	dB ← − + →	Voltage or current ratio	Power ratio
0·159	0·0251	16	6·31	39·8
0·126	0·0159	18	7·94	63·1
0·100	0·0100	20	10·00	100·0
$3{\cdot}16 \times 10^{-2}$	10^{-3}	30	$3{\cdot}16 \times 10$	10^3
10^{-2}	10^{-4}	40	10^2	10^4
$3{\cdot}16 \times 10^{-3}$	10^{-5}	50	$3{\cdot}16 \times 10^2$	10^5
10^{-3}	10^{-6}	60	10^3	10^6
$3{\cdot}16 \times 10^{-4}$	10^{-7}	70	$3{\cdot}16 \times 10^3$	10^7
10^{-4}	10^{-8}	80	10^4	10^8
$3{\cdot}16 \times 10^{-5}$	10^{-9}	90	$3{\cdot}16 \times 10^4$	10^9
10^{-5}	10^{-10}	100	10^5	10^{10}
$3{\cdot}16 \times 10^{-6}$	10^{-11}	110	$3{\cdot}16 \times 10^5$	10^{11}
10^{-6}	10^{-12}	120	10^6	10^{12}

Laws

Ampere's Rule Refers to the deflection direction of a magnetic pointer that is influenced by a current; an analogy being that if a person is assumed to be swimming with the current and facing the indicator, the north-seeking pole is deflected towards the left hand, the south pole being deflected in an opposite direction.

Ampere's Theorem The magnetic field from current flowing in a circuit is equivalent to that due to a simple magnetic shell, the outer edge coinciding with the electrical conductor with such strength that it equals that current strength.

Baur's Constant That voltage necessary to cause a discharge through a determined insulating material 1 mm thick. The law of dielectric strength is that breakdown voltage necessary to cause a discharge through a substance proportional to a 2/3 power of its thickness.

Coulomb's Law Implies that the mechanical force between two charged bodies is directly proportionate to the charges and inversely so to the squares of the distance separating them.

Faraday's Laws That of induction is that the e.m.f. induced in a circuit is proportional to the rate of change in the lines of force linking it. That of electrolysis is (1) That the quantity of a substance deposited in defined time is proportional to the current. (2) That different substances and quantities deposited by a single current in a similar time are proportional to the electro-chemical equivalents. The Faraday Effect states that when a light beam passes through a strong magnetic field the plane of polarisation is rotated.

Fleming's Rules By placing the thumb and first two fingers at right-angles respectively, the forefinger can represent the direction of magnetic field; the second finger, current direction; the thumb, motion direction. Use of the right hand in this way represents the relation in a dynamo; use of the left hand represents the relation in a motor.

Hall Effect If an electric current flows across the lines of flux of a magnetic field, an e.m.f. is observed at right-angles to the primary current and to the magnetic field. When a steady current flows in a magnetic field, e.m.f. tendencies develop at right-angles to the magnetic force and to the current, proportionately to the product of the current strength, the magnetic force and the sine of the angle between the direction of quantities.

Joule's Law As a formula this is I^2Rt joules. It refers to that heat developed by the current (I) which is proportional to the square of I multiplied by R and t, letting R = resistance and t = time. If the formula is seen as $JH = RI^2t$ it equals EIt, letting J = joules equivalent of heat, and H = the number of heat units.

Kerr Effect Illustrates that an angle of rotation is proportional to a magnetisation intensity and applies to the rotation of polarisation plane of plain polarised light as reflected from the pole of a magnet. The number (a constant) varies for different wavelengths and materials, making necessary the multiplication of magnetisation intensity in order to find the angle of rotation forming the effect.

Lenz's Law That induced currents have such a direction that the reaction forces generated have a tendency to oppose the motion or action producing them.

Maxwell's Law (*a*) Any two circuits carrying current tend so to dispose themselves as to include the largest possible number of lines of force common to the two. (*b*) Every electro-magnetic system tends to change its configuration so that the exciting circuit embraces the largest number of lines of force in a positive direction.

Maxwell's Rule Maxwell's *unit tubes* of electric or magnetic induction are such that a *unit pole* delivers 4π unit tubes of force.

Miller Circuit A form of circuit in which the time-constant of a resistance-capacitance combination is multiplied by means of the Miller effect on the capacitance. Named after John M. Miller.

Miller Effect Implies that the grid input impedance of a valve with a load in the anode circuit is different from its input impedance with a zero anode load. Should the load in the anode be resistance, the input impedance is purely capacitive. If the load impedance has a reactive component, the input impedance will have a resistive component. In pre-detector amplification, with a.v.c. to signal grids, the capacity across the tuned grid circuits tends to vary with the signal strength, evidencing detuning, the effect causing a charge (electrostatic) to be induced by the anode on the grid.

Planck's Constant Quanta of energy radiated when atomic electrons transfer from one state to another, assuming both to be *energy states* with electro-magnetic radiation. The constant (h) is given the value of $6{\cdot}626 \times 10^{-34}$ joule second. h is usually coupled to the symbol (ν) to represent the frequency of the radiated energy in hertz. That is, the frequency of the radiated energy is determinable by the relation $W_1 - W_2$, this equalling $h\nu$. W_1 and W_2 equal the values of the internal energy of the atom in initial and final stages. This constant is also known as the *Quantum Theory*.

Thévenin's Theorem The current through a resistance R connected across any two points A and B of an active network (i.e. a network containing one or more sources of e.m.f.) is obtained by dividing the p.d. between A and B, with R disconnected, by $(R+r)$, where r is the resistance of the network measured between points A and B with R disconnected and the sources of e.m.f. replaced by their internal resistances.

Connectors and connections

Data interchange by modems

When transmitting and receiving data across telephone or other circuits, the equipment which actually generates and uses the data (e.g., a computer or VDU terminal) is known as *data terminating equipment* (DTE). The equipment which terminates the telephone line and converts the basic data signals into signals which can be transmitted is known as *data circuit-terminating equipment* (DCE). As far as the user is concerned the interface between DTE and DCE is the most important. CCITT recommendation V24 defines the signal interchanges and functions between DTE and DCE; these are commonly known as the 100 series interchange circuits:

Interchange circuit		Data		Control		Timing	
Number	Name	From DCE	To DCE	From DCE	To DCE	From DCE	To DCE
101	Protective ground or earth						
102	Signal ground or common return						
103	Transmitted data		●				
104	Received data	●					
105	Request to send				●		
106	Ready for sending			●			
107	Data set ready			●			
108/1	Connect data set to line				●		
108/2	Data terminal ready				●		
109	Data channel received line signal detector			●			
110	Signal quality detector			●			
111	Data signalling rate selector (DTE)				●		
112	Data signalling rate selector (DCE)			●			
113	Transmitter signal element timing (DTE)						●
114	Transmitter signal element timing (DCE)					●	
115	Receiver signal element timing (DCE)					●	
116	Select standby				●		
117	Standby indicator			●			
118	Transmitted backward channel data		●				
119	Received backward channel data	●					
120	Transmit backward channel line signal				●		
121	Backward channel ready			●			
122	Backward channel received line signal detector			●			
123	Backward channel signal quality detector			●			
124	Select frequency groups				●		
125	Calling indicator			●			
126	Select transmit frequency				●		
127	Select receive frequency				●		
128	Receiver signal element timing (DTE)						●
129	Request to receive				●		
130	Transmit backward tone				●		
131	Received character timing					●	
132	Return to non-data mode				●		
133	Ready for receiving				●		
134	Received data present			●			
191	Transmitted voice answer				●		
192	Received voice answer			●			

Modem connector pin numbers

The connectors used with 100 series interchange circuits and its pin assignments are defined by international standard ISO 2110 and are (for modems following the CCITT recommendations V21, V23, V26, V26bis, V27 and V27bis) as follows:

Pin number	Interchange circuit numbers		
	V21	V23	V26/V27
1	*1	*1	*1
2	103	103	103
3	104	104	104
4	105	105	105
5	106	106	106
6	107	107	107
7	102	102	102
8	109	109	109
9	*N	*N	*N
10	*N	*N	*N
11	126	*N	*N
12	*F	122	122
13	*F	121	121
14	*F	118	118
15	*F	*2	114
16	*F	119	119
17	*F	*2	115
18	141	141	141
19	*F	120	120
20	108/1–2	108/1–2	108/1–2
21	140	140	140
22	125	125	125
23	*N	111	111
24	*N	*N	113
25	142	142	142

Notes:

*1 Pin 1 is assigned for connecting the shields between tandem sections of shielded cables. It may be connected to protective ground or signal ground.

*F Reserved for future use.

*N Reserved for national use.

Audio connectors

The DIN standards devised by the German Industrial Standards Board are widely used for the connection of audio equipment. The connectors are shown below. The 3-way and 5-way 45° are the most common, and connections for those are listed.

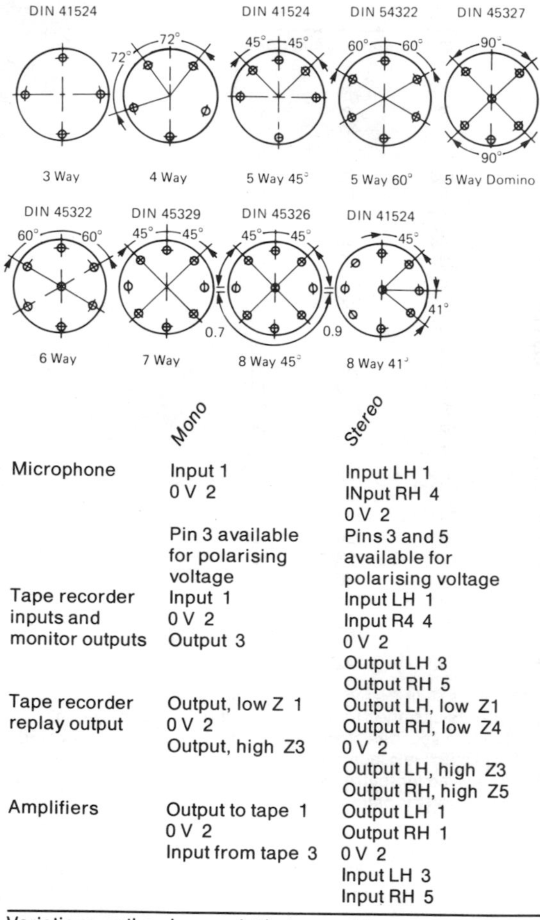

	Mono	Stereo
Microphone	Input 1 0 V 2 Pin 3 available for polarising voltage	Input LH 1 INput RH 4 0 V 2 Pins 3 and 5 available for polarising voltage
Tape recorder inputs and monitor outputs	Input 1 0 V 2 Output 3	Input LH 1 Input R4 4 0 V 2 Output LH 3 Output RH 5
Tape recorder replay output	Output, low Z 1 0 V 2 Output, high Z3	Output LH, low Z1 Output RH, low Z4 0 V 2 Output LH, high Z3 Output RH, high Z5
Amplifiers	Output to tape 1 0 V 2 Input from tape 3	Output LH 1 Output RH 1 0 V 2 Input LH 3 Input RH 5

Variations on the above exist between different manufacturers.

Coaxial connectors

A number of connectors are available for use with coaxial cables. The most common are:

Type	Impedance (Ω)	Maximum fre. (MHz)	Max. peak voltage (V)	Notes
BNC	50 75	10 000 10 000	500 500	High quality constant impedance bayonet fitting connector.
Miniature BNC	50	10 000	500	High quality constant impedance connector.
N	50 75	10 000 10 000	1000 500	Recommended for high power circuits above 400 MHz.
PL259/ SO239	50	200	500	Non-constant impedance design. High vswr makes it unsuitable for use above 144 MHz.
C	75	—	—	Bayonet fitting
F	50	—	—	American cctv connector used on some 144 MHz portable transceivers. Plugs use inner conductor of cable for centre pin.
Belling Lee	75	—	—	British tv aerial connector used extensively on home-built equipment and some British commercial equipment. Main virtue is low cost but aluminium plugs can corrode when used outside.
Phono	—	—	—	American connector originally designed for audio use.
GR	50	1000	—	Constant impedance sexless connectors.

RF Cables USA RG series

RG number	Nominal impedance Z_o (ohms)	Overall diameter (inches)	Velocity factor	Attenuation (dB per 100 ft) 1 MHz	10 MH₂	10 MHz	1000MHz	3000 MHz	Capacity (pF/ft)	Maximum operating voltage RMS
RG-5/U	52·5	0·332	0·659	0·21	0·77	2·9	11·5	22·0	28·5	3000
RG-5B/U	50·0	0·332	0·659	0·16	0·66	2·4	8·8	16·7	29·5	3000
RG-6A/U	75·0	0·332	0·659	0·21	0·78	2·9	11·2	21·0	20·0	2700
RG-8A/U	50·0	0·405	0·659	0·16	0·55	2·0	8·0	16·5	30·5	4000
RG-9/U	51·0	0·420	0·659	0·16	0·57	2·0	7·3	15·5	30·0	4000
RG-9B/U	50·0	0·425	0·659	0·175	0·61	2·1	9·0	18·0	30·5	000
RG-10A/U	50·0	0·475	0·659	0·16	0·55	2·0	8·0	16·5	30·5	4000
RG-11A/U	75·0	0·405	0·66	0·18	0·7	2·3	7·8	16·5	20·5	5000
RG-12A/U	75·0	0·475	0·659	0·18	0·66	2·3	8·0	16·5	20·5	4000
RG-13A/U	75·0	0·425	0·659	0·18	0·66	2·3	8·0	16·5	20·5	4000
RG-14A/U	50·0	0·545	0·659	0·12	0·41	1·4	5·5	12·0	30·0	5500
RG-16/U	52·0	0·630	0·670	0·1	0·4	1·2	6·7	16·0	29·5	6000
RG-17A/U	50·0	0·870	0·659	0·066	0·225	0·80	3·4	8·5	30·0	11000
RG-18A/U	50·0	0·945	0·659	0·066	0·225	0·80	3·4	8·5	30·5	11000

RG-19A/U	50·0	1·120	0·659	0·04	0·17	0·68	3·5	7·7	30·5	14000
RG-20A/U	50·0	1·195	0·659	0·04	0·17	0·68	3·5	7·7	30·5	14000
RG-21/AU	50·0	0·332	0·659	1·4	4·4	13·0	43·0	85·0	30·0	2700
RG-29/U	53·5	0·184	0·659	0·33	1·2	4·4	16·0	30·0	28·5	1900
RG-34A/U	75·0	0·630	0·659	0·065	0·29	1·3	6·0	12·5	20·5	5200
RG-34B/U	75	0·630	0·66		0·3	1·4	5·8		21·5	6500
RG-35A/U	75·0	0·945	0·659	0·07	0·235	0·85	3·5	8·60	20·5	10000
RG-54A/U	58·0	0·250	0·659	0·18	0·74	3·1	11·5	21·5	26·5	3000
RG-55/U	53·5	0·206	0·659	0·36	1·3	4·8	17·0	32·0	28·5	1900
RG-55A/U	50·0	0·216	0·659	0·36	1·3	4·8	17·0	32·0	29·5	1900
RG-58/U	53·5	0·195	0·659	0·33	1·25	4·65	17·5	37·5	28·5	1900
RG-58C/U	50·0	0·195	0·659	0·42	1·4	4·9	24·0	45·0	30·0	1900
RG-59A/U	75·0	0·242	0·659	0·34	1·10	3·40	12·0	26·0	20·5	2300
RG-59B/U	75	0·242	0·66		1·1	3·4	12		21	2300
RG-62A/U	93·0	0·242	0·84	0·25	0·85	2·70	8·6	18·5	13·5	750
RG-74A/U	50·0	0·615	0·659	0·10	0·38	1·5	6·0	11·5	30·0	5500
RG-83/U	35·0	0·405	0·66	0·23	0·80	2·8	9·6	24·0	44·0	2000
*RG-213/U	50	0·405	0·66	0·16	0·6	1·9	8·0		29·5	5000
†RG-218/U	50	0·870	0·66	0·066	0·2	1·0	4·4		29·5	11000
‡RG-220/U	50	1·120	0·66	0·04	0·2	0·7	3·6		29·5	14000

*Formerly RG8A/U.
†Formerly RG17A/U.
‡Formerly RG19A/U.

British UR series

UR number	Nominal impedance Z_0 (ohms)	Overall diameter (inches)	Inner conductor (inches)	Capacity (pF/ft)	Maximum operating voltage RMS	Attenuation (dB per 100 ft)				Nearest RG equivalent
						10 MHz	100 MHz	300 MHz	1000 MHz	
43	52	0·195	0·032	29	2750	1·3	4·3	8·7	18·1	58/U
57	75	0·405	0·044	20·6	5000	0·6	1·9	3·5	7·1	11A/U
63*	75	0·855	0·175	14	4400	0·15	0·5	0·9	1·7	
67	50	0·405	7/0·029	30	4800	0·6	2·0	3·7	7·5	213/U
74	51	0·870	0·188	30·7	15000	0·3	1·0	1·9	4·2	218/U
76	51	0·195	19/0·0066	29	1800	1·6	5·3	9·6	22·0	58C/U
77	75	0·870	0·104	20·5	125000	0·3	1·0	1·9	4·2	164/U
79*	50	0·855	0·265	21	6000	0·16	0·5	0·9	1·8	
83*	50	0·555	0·168	21	2600	0·25	0·8	1·5	2·8	
85*	75	0·555	0·109	14	2600	0·2	0·7	1·3	2·5	
90	75	0·242	0·022	20	2500	1·1	3·5	6·3	12·3	59B/U

All the above cables have solid dielectric with a velocity factor of 0·66 with the exception of those marked with an asterisk, which are helical membrane and have a velocity factor of 0·96.

	USA size	Nominal voltage (V)	Type†	IEC equivalent	Maximum dimensions (mm): Length (or diameter)	Width	Height	Contacts	Current (mA)	Weight (g)
Zinc carbon	N	1·5	D23	R1	12	—	30·1	Cap and base	1–5	7
	AAA		HP16	R03	10·5	—	45	Cap and base	0–10000	8·5
	AA		HP7	R6	14·5	—	50·5	Cap and base	0–75	16·5
	AA		C7	R6	14·5	—	50·5	Cap and base	0–75	16·5
	C		SP11	R14	26·2	—	50	Cap and base	20–60	45
	C		HP11	R14	26·2	—	50	Cap and base	0–1000	45
	C		C11	R14	26·2	—	50	Cap and base	0–5	45
	D		SP2	R20	34·2	—	61·8	Cap and base	25–100	90
	D		HP2	R20	34·2	—	61·8	Cap and base	0–2000	90
		4·5	AD28	3R25	101·6	34·9	106	Socket	30–300	453·6
			1289	3R12	62	22	67	Flat springs	0–300	113
		6·0	PP8	4-F100-4	65·1	51·6	200·8	Press studs	20–151	1100
			PJ996	4-R25	67	67	102	Spiral springs	30–300	581
			991		135·7	72·2	125·4	Two screws	30–500	1470
		9·0	PP3-P	6-F22	26·5	17·5	48·5	Press studs	0–50	39
			PP3-C	6-F22	26·5	17·5	48·5	Press studs	0–50	39
			PP3	6-F22	26·5	17·5	48·5	Press studs	0–10	38
			PP4	6-F20	25·5	—	50	Press studs	0–10	51
			PP6	6-F50-2	36	34·5	70	Press studs	2·5–15	142
			PP7	6-F90	46	46	61·9	Press studs	5–20	198
			PP9	6-F100	66	52	81	Press studs	5–50	425
			PP10	6-F100-3	66	52	226	Socket	15–150	1250
		15·0	B154	10-F15	16	15	35	End contacts	0·1–0·5	14·2
			B121	10-F20	27	16	37	End contacts	0·1–1·0	21
		22·5	B155	15-F15	16	15	51	End contacts	0·1–0·5	20
			B122	15-F20	27	16	51	End Contacts	0·1–1·0	32

	USA size	Nominal voltage (V)	Type‡	IEC equivalent	Maximum dimensions (mm) Length (or diameter)	Width	Height	Contacts	Current (mA)	Weight (g)
Manganese alkaline	ED	1·5	MN1300*	LR20	34·2	—	61·5	Cap and base	10·00†	125
	C		MN1400*	LR14	26·2	—	50	Cap and base	5·50†	65
	AA		MN1500*	LR6	14·5	—	50·5	Cap and base	1·80†	23
	AAA		MN2400*	LR03	10·5	—	44·5	Cap and base	0·80†	13
	N		MN9100*	LR1	12	—	30·2	Cap and base	0·65†	9·6
Mercuric oxide		1·35/1·4	RM675H	NR07	11·6	—	5·4	Cap and base (button)	0·21†	2·6
			RM625N	MR9	15·6	—	6·2	Cap and base (button)	0·25†	4·3
			RM575H	NR08	11·6	—	3·5	Cap and base (button)	0·12†	1·4
			RM1H	NR50	16·4	—	16·8	Cap and base (button)	1·00†	12·0
Silver oxide		1·5	10L14	5R44	11·56	—	5·33	Cap and base (button)	0·13†	2·2
			10L124	5R43	11·56	—	4·19	Cap and base (button)	0·13†	1·7
			10L123	5R48	7·75	—	5·33	Cap and base (button)	0·08†	1·0
			10L125	5R41	7·75	—	3·58	Cap and base (button)	0·04†	0·8
Nickel cadmium		1·25	NC828	—				Button	0·28†	16·5
	AA		NCC60	—	See HP7			Button	0·60†	30·0
	C		NCC200	—	See HP11			Button	2·00†	78·0
	D		NCC400	—	See HP2			Button	4·00†	170·0
		10·0	NC828/8	—				Button stack	0·28†	126·0
		12·0	10/225DK	—				Button stack	0·225†	135·0
		9·0	TR7/8	(DEAC)	See PP3			Press studs	0·07†	45·0
	AA	1·25	501RS	(DEAC)	See HP7			Press studs	0·50†	30·0
	C		RS1·8	(DEAC)	See HP11			Press studs	1·80†	65·0
	D		RS4	(DEAC)	See HP2			Press studs	4·00†	150·0

†Capacity in ampere hours.
‡BEREC types unless otherwise indicated.

Powers of 2

n	2^n	2^{-n}	$-n$
1	2	·5	−1
2	4	·25	−2
3	8	·125	−3
4	16	·062 5	−4
5	32	·031 25	−5
6	64	·015 625	−6
7	128	·007 812 5	−7
8	256	·003 906 25	−8
9	512	·001 953 125	−9
10	1 024	·000 976 562 5	−10
11	2 048	·000 488 281 25	−11
12	4 096	·000 244 140 625	−12
13	8 192	·000 122 070 312 5	−13
14	16 384	·000 061 035 156 25	−14
15	32 768	·000 030 517 578 125	−15
16	65 536	·000 015 258 789 062 5	−16
17	131 072	·000 007 629 394 531 25	−17
18	262 144	·000 003 814 679 265 625	−18
19	524 288	·000 001 907 348 632 812 5	−19
20	1 048 576	·000 000 953 674 316 406 25	−20
21	2 097 152	·000 000 476 837 158 203 125	−21
22	4 194 304	·000 000 238 418 579 101 562 5	−22
23	8 388 608	·000 000 119 209 289 550 781 25	−23
24	16 777 216	·000 000 059 604 644 775 390 625	−24
25	33 554 432	·000 000 029 802 322 387 695 312 5	−25
26	67 108 864	·000 000 014 901 161 193 847 656 25	−26
27	134 217 728	·000 000 007 450 580 596 923 828 125	−27
28	268 435 456	·000 000 003 725 290 298 461 914 062 5	−28
29	536 870 912	·000 000 001 862 645 149 230 957 031 25	−29
30	1 073 741 824	·000 000 000 931 322 574 615 478 515 625	−30
31	2 147 483 648	·000 000 000 465 661 287 307 739 257 812 5	−31
32	4 294 967 296	·000 000 000 232 830 643 653 869 628 906 25	−32

Powers of 10_{16}

			10^{n}	n	10^{-n}					
			1	0	1·0000	0000	0000	0000		
			A	1	0·1999	9999	9999	999A		
			64	2	0·28F5	C28F	5C28	F5C3	×	16^{-1}
			3E8	3	0·4189	374B	C6A7	EF9E	×	16^{-2}
			2710	4	0·68DB	8BAC	710C	B296	×	16^{-3}
		1	86A0	5	0·A7C5	AC47	1B47	8423	×	16^{-4}
		F	4240	6	0·10C6	F7A0	B5ED	8D37	×	16^{-4}
		98	9680	7	0·1AD7	F29A	BCAF	4858	×	16^{-5}
		5F5	E100	8	0·2AF3	1DC4	6118	73BF	×	16^{-6}
		3B9A	CA00	9	0·44B8	2FA0	9B5A	52CC	×	16^{-7}
	2	540B	E400	10	0·6DF3	7F67	5EF6	EADF	×	16^{-8}
	17	4876	E800	11	0·AFEB	FF0B	CB24	AAFF	×	16^{-9}
	E8	D4A5	1000	12	0·1197	9981	2DEA	1119	×	16^{-9}
	918	4E72	A000	13	0·1C25	C268	4976	81C2	×	16^{-10}
	5AF3	107A	4000	14	0·2D09	370D	4257	3604	×	16^{-11}
3	8D7E	A4C6	8000	15	0·480E	BE7B	9D58	566D	×	16^{-12}
23	86F2	6FC1	0000	16	0·734A	CA5F	6226	F0AE	×	$16 \times^{-13}$
163	4578	5D8A	0000	17	0·B877	AA32	36A4	B449	×	16^{-14}
DE0	B6B3	A764	0000	18	0·1272	5DD1	D243	ABA1	×	16^{-14}
8AC7	2304	89E8	0000	19	0·1D83	C94F	B6D2	AC35	×	16^{-15}

Powers of 16_{10}

16^n	n	16^{-n}					
1	0	0·10000	00000	00000	00000	×	10
16	1	0·62500	00000	00000	00000	×	10^{-1}
256	2	0·39062	50000	00000	00000	×	10^{-2}
4 096	3	0·24414	06250	00000	00000	×	10^{-3}
65 536	4	0·15258	78906	25000	00000	×	10^{-4}
1 048 576	5	0·95367	43164	06250	00000	×	10^{-6}
16 777 216	6	0·59604	64477	53906	25000	×	10^{-7}
268 435 456	7	0·37252	90298	46191	40625	×	10^{-8}
4 294 967 296	8	0·23283	06436	53869	62891	×	10^{-9}
68 719 476 736	9	0·14551	91522	83668	51807	×	10^{-10}
1 099 511 627 776	10	0·90949	47017	72928	23792	×	10^{-12}
17 592 186 044 416	11	0·56843	41886	08080	14870	×	10^{-13}
281 474 976 710 656	12	0·35527	13678	80050	09294	×	10^{-14}
4 503 599 627 370 496	13	0·22204	46049	25031	30808	×	10^{-15}
72 057 594 037 927 936	14	0·13877	78780	78144	56755	×	10^{-16}
1 152 921 504 606 846 976	15	0·86736	17379	88403	54721	×	10^{-18}

Sounds and sound levels

Sound pressure (mPa)	Pressure ratio	Intensity ratio	Sound level (dB)	Source or description of typical sound
0·2 (datum)	1 $(=10^{0})$	1	0	Sound-proof room (threshold of hearing)
0·063	3·16 $(=10^{0\cdot5})$	10^{1}	10	Rustle of leaves in a breeze
0·2	10 $(=10^{1})$	10^{2}	20	Whisper
0·63	31·6 $(=10^{1\cdot5})$	10^{3}	30	Quiet conversation
2	100 $(=10^{2})$	10^{4}	40	Suburban home
6·3	316 $(=10^{2\cdot5})$	10^{5}	50	Typical conversation
20	1000 $(=10^{3})$	10^{6}	60	Large shop
63	3160 $(=10^{3\cdot5})$	10^{7}	70	City street
200	10000 $(=10^{4})$	10^{8}	80	Noisy office with typing
630	31600 $(=10^{4\cdot5})$	10^{9}	90	Underground railway
2000	100000 $(=10^{5})$	10^{10}	100	Pneumatic drill at 3 m
6300	316000 $(=10^{5\cdot5})$	10^{11}	110	Prop aircraft taking off
20000	1000000 $(=10^{6})$	10^{12}	120	Jet aircraft taking off (threshold of pain)

Celsius–Fahrenheit conversion table

C	F	C	F	C	F	C	F
0	32	265	509	530	986	795	1,463
5	41	270	518	535	995	800	1,472
10	50	275	527	540	1,004	805	1,481
15	59	280	536	545	1,013	810	1,490
20	68	285	545	550	1,022	815	1,499
25	77	290	554	555	1,031	820	1,508
30	86	295	563	560	1,040	825	1,517
35	93	300	572	565	1,049	830	1,526
40	104	305	581	570	1,058	835	1,535
45	113	310	590	575	1,067	840	1,544
50	122	315	599	580	1,076	845	1,553
55	131	320	608	585	1,085	850	1,562
60	140	325	617	590	1,094	855	1,571
65	149	330	626	595	1,103	860	1,580
70	158	335	635	600	1,112	865	1,589
75	167	340	644	605	1,121	870	1,598
80	176	345	653	610	1,130	875	1,607
85	185	350	662	615	1,139	880	1,616
90	194	355	671	620	1,148	885	1,625
95	203	360	680	625	1,157	890	1,634
100	212	365	689	630	1,166	895	1,643
105	221	370	698	635	1,175	900	1,652
110	230	375	707	640	1,184	905	1,661
115	239	380	716	645	1,193	910	1,670
120	248	385	725	650	1,202	915	1,679
125	257	390	734	655	1,211	920	1,688

130	266	395	743	660	1,220	925	1,697
135	275	400	752	665	1,229	930	1,706
140	284	405	761	670	1,238	935	1,715
145	293	410	770	675	1,247	940	1,724
150	302	415	779	680	1,256	945	1,733
155	311	420	788	685	1,265	950	1,742
160	320	425	797	690	1,274	955	1,751
165	329	430	806	695	1,283	960	1,760
170	338	435	815	700	1,292	965	1,769
175	347	440	824	705	1,301	970	1,778
180	356	445	833	710	1,310	975	1,787
185	365	450	842	715	1,319	980	1,796
190	374	455	851	720	1,328	985	1,805
195	383	460	860	725	1,337	990	1,814
200	392	465	869	730	1,346	995	1,823
205	401	470	877	735	1,355	1,000	1,832
210	410	475	887	740	1,364	1,005	1,841
215	419	480	896	745	1,373	1,010	1,850
220	428	485	905	750	1,382	1,015	1,859
225	437	490	914	755	1,391		
230	446	495	923	760	1,400		
235	455	500	932	765	1,409		
240	464	505	941	770	1,418		
245	473	510	950	775	1,427		
250	482	515	959	780	1,436		
255	491	520	968	785	1,445		
260	500	525	977	790	1,454		

Temperature conversion formulae

°F to °C °C = 5/9 (°F − 32)
°C to °F °F = 9/5 °C + 32
°F to °R °R = 4/9 (°F − 32)
°R to °F °F = 9/4 °R + 32
°R to °C °C = 5/4 °R
Absolute zero = −273·14°C.

Statistical formulae

The **arithmetic mean** of a set of numbers $X_1, X_2, \ldots, X_N$ is their average. It is the sum of the numbers divided by the number of numbers and is denoted by $\bar{X}$

$$\bar{X} = \frac{X_1 + X_2 + X_3 \ldots X_N}{N} = \frac{\sum_{i=1}^{N} X_1}{N}$$

The **standard deviation** is denoted by σ.

$$\sigma = \sqrt{\frac{\text{sum of squares of differences between numbers and mean}}{N}}$$

$$= \sqrt{\frac{\sum_{I=1}^{N} (X_I - \bar{X})^2}{N}}$$

Calculus

Differentiation

The derivative of a function $y = f(t)$ is denoted by

$\frac{dy}{dt}$ or $\dot{y}$ if t represents time

The second derivative of $y = f(t)$ is denoted by $\frac{d^2y}{dt^2}$ or $\ddot{y}$ if t is time.

Useful derivatives

function $y = f(t)$		derivative $\frac{dy}{dt}$
1		0
t		1
t^A	$(A \neq 0)$	At^{A-1}
$\sin \omega t$	$(\omega \neq 0)$	$\omega \cos \omega t$
$\cos \omega t$	$(\omega \neq 0)$	$-\omega \sin \omega t$
$\tan at$	$(a \neq 0)$	$a \sec^2 at$
$\exp at$	$(a \neq 0)$	$a \exp at$
$\log_e at$	$(a \neq 0)$	$\frac{1}{t}$

Standard integrals

function $f(t)$	standard integrals $\int f(t)\,dt$
1	t
t	$\frac{1}{2}t^2$
$t^N (N \neq -1)$	$\frac{1}{N+1} t^{N+1} \quad (N \neq -1)$
$\frac{1}{T}$	$\log_e T \quad (T > 0)$
$\sin \omega t$	$-\frac{1}{\omega} \cos \omega t \quad (\omega \neq 0)$
$\cos \omega t$	$\frac{1}{\omega} \sin \omega t \quad (\omega \neq 0)$
$\exp at \quad (a \neq 0)$	$\frac{1}{a} \exp at \quad (a \neq 0)$
$\frac{1}{a^2 - t^2}$	$\frac{1}{2a} \log_e \left(\frac{a+t}{a-t}\right) (-a < t < +a)$
$\log_e(at)$	$t[\log_e(at) - 1]$

Mensuration

A and a = area; b = base; C and c = circumference; D and d = diameter; h = height; $n°$ = number of degrees; p = perpendicular; R and r = radius; s = span or chord; v = versed sine.

Square: $a = \text{side}^2$; side $= \sqrt{a}$;
diagonal = side $\times \sqrt{2}$.

Rectangle or parallelogram: $a = bp$.

Trapezoid (two sides parallel): a = mean length parallel sides × distance between them.

Triangle: $a = \frac{1}{2}bp$

Irregular figure: a = weight of template ÷ weight of square inch of similar material.

Side of square multiplied by 1·4142 equals diameter of its circumscribing circle.

A side multiplied by 4·443 equals circumference of its circumscribing circle.

A side multiplied by 1·128 equals diameter of a circle of equal area.

Circle: $a = \pi r^2 = d^2\pi/4 = 0{\cdot}7854d^2 = 0{\cdot}5\,cr$; $c = 2\pi r = d\pi = 3{\cdot}1416d = 3{\cdot}54\sqrt{a}$ = (approx.) $\frac{22}{7}d$. Side of equal square = $0{\cdot}8862d$; side of inscribed square = $0{\cdot}7071d$; $d = 0{\cdot}3183c$. A circle has the maximum area for a given perimeter.

Annulus of circle: $a = (D + d)(D - d)\,\frac{\pi}{4}$

$$= (D^2 - d^2)\,\frac{\pi}{4}$$

Segment of circle:

a = area of sector − area of triangle

$$= \frac{4v}{3}\sqrt{(0{\cdot}625v)^2 + (\tfrac{1}{2}S)^2}.$$

Length of arc = $0{\cdot}0174533n°r$; length of

$$\text{arc} = \tfrac{1}{3}\left(8\sqrt{\frac{S^2}{4} + v^2 - s}\right);$$

approx. length of arc = ⅓ (8 times chord of ½ arc − chord of whole arc).

$$d = \frac{(\tfrac{1}{2}\text{ chord})}{v} + v;$$

$$\text{radius of curve} = \frac{S^2}{8V} + \frac{V}{2}.$$

Sector of circle: $a = 0{\cdot}5r$ × length arc;

$= n°$ × area circle ÷ 360.

Ellipse: $a = \frac{\pi}{4}Dd = \pi Rr$; c (approx.)

$$= \sqrt{\frac{D^2 + d^2}{2}} \times \pi;\ c\text{ (approx.)} = \pi\frac{Da}{2}.$$

Parabola: $a = \frac{2}{3}bh$.
Cone or pyramid: surface

$$= \frac{\text{circ. of base} \times \text{slant length}}{2} + \text{base};$$

contents = area of base × ⅓ vertical height.

Frustrum of cone:
surface = $(C + c) \times \frac{1}{2}$ slant height + ends;
contents = $0{\cdot}2618h(D^2 + d^2 + Dd)$;

$$= \tfrac{1}{3}h(A + a + \sqrt{A \times a}).$$

Wedge: contents = $\frac{1}{6}$ (length of edge + 2 length of back)bh.
Oblique prism: contents = area base × height.
Sphere: surface $= d^2\pi = 4\pi r^2$,

$$\text{contents} = d^3\frac{\pi}{6} = \frac{4}{3}\pi r^3.$$

Segment of sphere: r = rad. of base;

contents = $\dfrac{\pi}{6}h(3r^2 + h^2)$; r = rad. of sphere;

contents = $\dfrac{\pi}{3}h^2(3r - h)$.

Spherical zone:

contents = $\dfrac{\pi}{2}h(\tfrac{1}{3}h^2 + R^2 + r^2)$; surface of convex part of

segment or zone of sphere = πd(of sph.)$h = 2\pi rh$.

Mid. sph. zone: contents = $(r + \tfrac{2}{3}h^2)\dfrac{\pi}{4}$

Spheroid:

contents = revolving axis^2 × fixed axis × $\dfrac{\pi}{6}$.

Cube or rectangular solid contents = length × breadth × thickness.

Prismoidal formula: contents

$$= \frac{\text{end areas} + 4 \text{ times mid. area} \times \text{length}}{6}$$

Solid revolution: contents = a of generating plane × c described by centroid of this plane during revolution. Areas of similar plane figures are as the squares of like sides. Contents of similar solids are as the cubes of like sides.

Rules relative to the circle, square, cylinder, etc.:
To find circumference of a circle:
Multiply diameter by 3·1416; or divide diameter by 0·3183.
To find diameter of a circle:
Multiply circumference by 0·3183; or divide circumference by 3·1416.
To find radius of a circle:
Multiply circumference by 0·15915; or divide circumference by 6·28318.
To find the side of an inscribed square:
Multiply diameter by 0·7071; or multiply circumference by 0·2251; or divide circumference by 4·4428.
To find side of an equal square:
Multiply diameter by 0·8862; or divide diameter by 1·1284; or multiply circumference by 0·2821; or divide circumference by 3·545.

To find area of a circle:

Multiply circumference by ¼ of the diameter; or multiply the square of diameter by 0·7854; or multiply the square of circumference by 0·07958; or multiply the square of ½ diameter by 3·1416.

To find the surface of a sphere or globe:

Multiply the diameter by the circumference; or multiply the square of diameter by 3·1416; or multiply 4 times the square of radius by 3·1416.

Cylinder.

To find the area of surface:

Multiply the diameter by 3⅐ × length.

Capacity – 3⅐ × radius^2 × height.

Values and Powers of:

$\pi = 3{\cdot}1415926536$, or $3{\cdot}1416$, or $^{22}/_7$ or $3^1/_7$;

$\pi^2 = 9{\cdot}86965$; $\sqrt{\pi} = 1{\cdot}772453$;

$$\frac{1}{\pi} = 0{\cdot}31831; \quad \frac{\pi}{2} = 1{\cdot}570796;$$

$$\frac{\pi}{3} = 1{\cdot}047197.$$

Radian = 57·2958 degrees.

Table A

Fig. 1. Diagram for Table A.

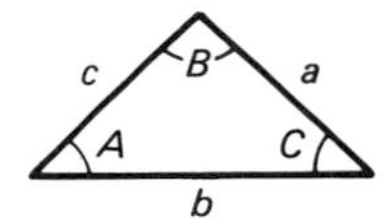

Parts given	Parts to be found	Formulae
abc	*A*	$\cos A = \dfrac{b^2 + c^2 - a^2}{2bc}$
abA	*B*	$\sin B = \dfrac{b \times \sin A}{a}$
abA	*C*	$C = 180° - (A + B)$
aAB	*b*	$b = \dfrac{a \times \sin B}{\sin A}$
aAB	*c*	$c = \dfrac{a \sin C}{\sin A} = \dfrac{a \sin (180° - A - B)}{\sin A}$
abC	*B*	$B = 180° - (A + C)$

Table B

Fig. 2. Diagram for Table B.

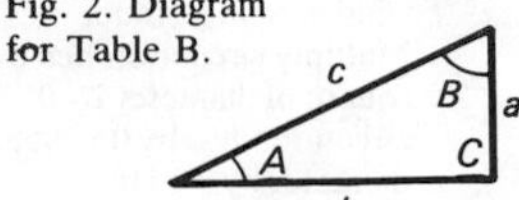

Parts given				
a & c	$\sin A = \frac{a}{c}$	$\cos B = \frac{a}{c}$	$b = \sqrt{c^2 - a^2}$	
a & b	$\tan A = \frac{a}{b}$	$\cot B = \frac{a}{b}$	$c = \sqrt{a^2 + b^2}$	
c & b	$\cos A = \frac{b}{c}$	$\sin B = \frac{b}{c}$	$a = \sqrt{c^2 - b^2}$	
A & a		$B = 90° - A$	$b = a \times \cot A$	$c = \frac{a}{\sin A}$
A & b		$B = 90° - A$	$a = b \times \tan A$	$c = \frac{b}{\cos A}$
A & c		$B = 90° - A$	$a = c \times \sin A$	$b = c \times \cos A$

Fig. 3. In any right-angled triangle:

$\tan A = \frac{BC}{AC}$, $\sin A = \frac{BC}{AB}$

$\cos A = \frac{AC}{AB}$, $\cot A = \frac{AC}{BC}$

$\sec A = \frac{AB}{AC}$, $\operatorname{cosec} A = \frac{AB}{BC}$

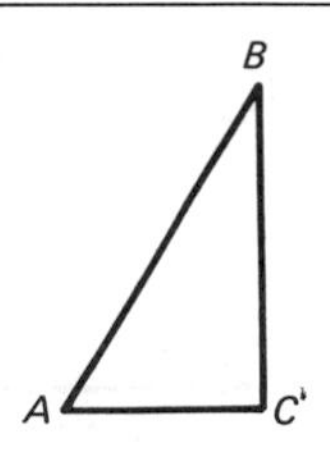

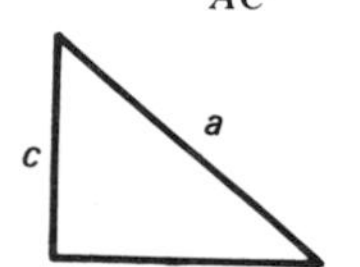

Fig. 4. In any right-angled triangle:

$$a^2 = c^2 + b^2$$
$$c = \sqrt{a^2 - b^2}$$
$$b = \sqrt{a^2 - c^2}$$
$$a = \sqrt{b^2 + c^2}$$

Fig. 5. $c + d : a + b :: b - a : d - c$.

$$d = \frac{c + d}{2} + \frac{d - c}{2}$$

$$x = \sqrt{b^2 - d^2}$$

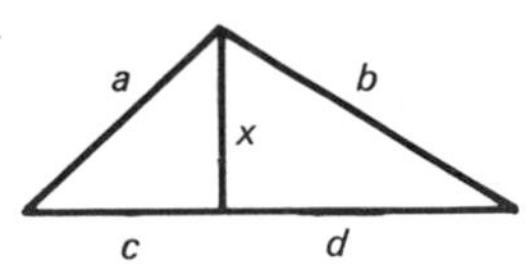

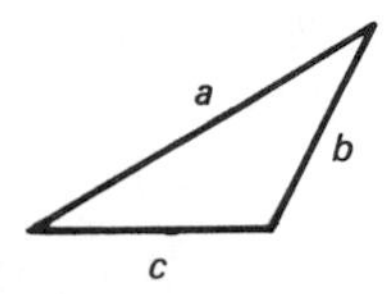

In Fig. 6, where the lengths of three sides only are known:
area = $\sqrt{s(s - a)(s - b)(s - c)}$
where $s = \frac{a + b + c}{2}$

Fig. 7. In this diagram:

$a:b::b:c$ or $\frac{b^2}{a} = c.$

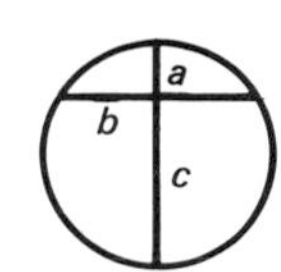

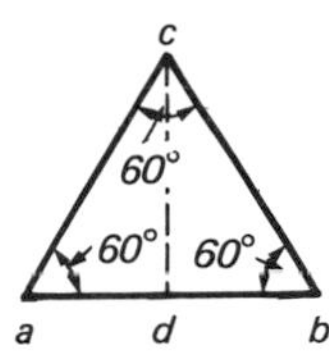

Fig. 8. In an equilateral triangle $ab = 1$, then $cd = \sqrt{0{\cdot}75} = 0{\cdot}866$, and $ad = 0{\cdot}5$; $ab = 2$, then $cd = \sqrt{3{\cdot}0} = 1{\cdot}732$, and $ad = 1$; $cd = 1$, then $ac = 1{\cdot}155$ and $ad = 0{\cdot}577$; $cd = 0{\cdot}5$, then $ac = 0{\cdot}577$ and $ad = 0{\cdot}288$.

Fig. 9. In a right-angled triangle with two equal acute angles, $bc = ac$, $bc = 1$, then $ab = \sqrt{2} = 1{\cdot}414$; $ab = 1$, then $bc = \sqrt{0{\cdot}5} = 0{\cdot}707$.

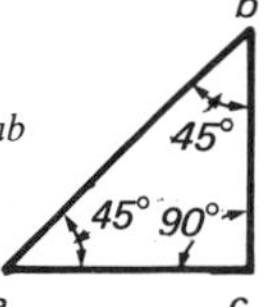

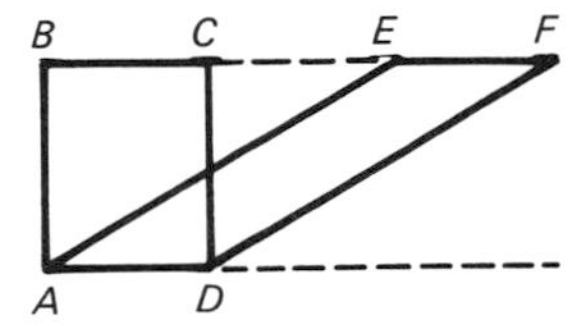

Fig. 10 shows that parallelograms on the same base and between the same parallels are equal; thus $ABCD = ADEF$.

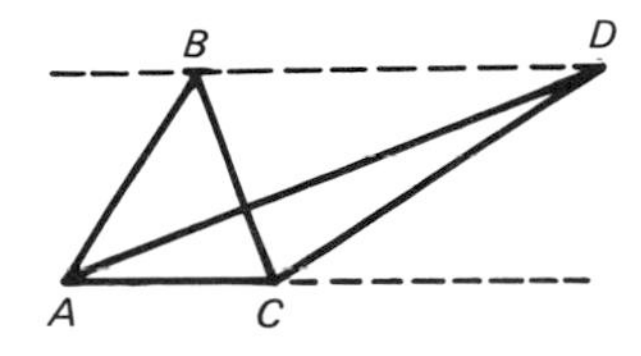

Fig. 11 demonstrates that triangles on the same base and between the same parallels are equal in area; thus, $ABC = ADC$.

Trigonometric relationships

$$\sin\left(\frac{\pi}{2} - \alpha\right) = \cos\alpha$$

$\sin(-\alpha) = -\sin\alpha$

$\sin(\pi - \alpha) = \sin\alpha$

$\sin(\pi + \alpha) = -\sin\alpha$

$\sin(2\pi - \alpha) = \sin(-\alpha) = -\sin\alpha$

$\sin(2N\pi + \alpha) = \sin\alpha$ (N an integer)

$$\frac{\sin\alpha}{\cos\alpha} = \tan\alpha$$

$$\cos\left(\frac{\pi}{2} = \alpha\right) = \sin\alpha$$

$\cos(-\alpha) = \cos\alpha$
$\cos(\pi - \alpha) = -\cos\alpha$
$\cos(\pi + \alpha) = -\cos\alpha$
$\cos(2\pi - \alpha) = \cos(-\alpha) = \cos\alpha$
$\cos(2\pi N + \alpha) = \cos\alpha$ (N and integer)

$$\tan\left(\frac{\pi}{2} - \alpha\right) = \frac{1}{\tan\alpha}$$

$\tan(-\alpha) = -\tan\alpha$
$\tan(\alpha + N\pi) = \tan\alpha$ (N an integer)
$\sin^2\alpha + \cos^2\alpha = 1$
$\sin^2\alpha = \tfrac{1}{2}(1 - \cos 2\alpha)$
$\cos^2\alpha = \tfrac{1}{2}(1 + \cos 2\alpha)$
$\tan^2\alpha + 1 = \sec^2\alpha$

$\sin(\alpha + \beta) = \sin\alpha\cos\beta + \sin\beta\cos\alpha$
$\cos(\alpha + \beta) = \cos\alpha\cos\beta - \sin\alpha\sin\beta$

$$\tan(\alpha + \beta) = \frac{\tan\alpha + \tan\beta}{1 - \tan\alpha\tan\beta}$$

(α,β can be positive or negative)

$\sin 2\alpha = 2\sin\alpha\cos\alpha$

$$\begin{aligned}\cos 2\alpha &= \cos^2\alpha - \sin^2\alpha \\ &= 2\cos^2\alpha - 1 \\ &= 1 - 2\sin^2\alpha\end{aligned}$$

$$\tan 2\alpha = \frac{2\tan\alpha}{1 - \tan^2\alpha} \qquad (\tan\alpha \neq \pm 1)$$

Transistor circuits and characteristics

Basic transistor circuits showing signal source and load (R_L)	Common base	Common emitter
	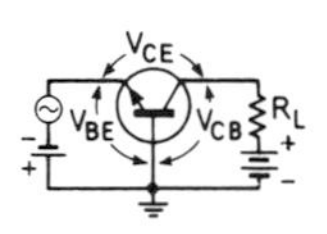	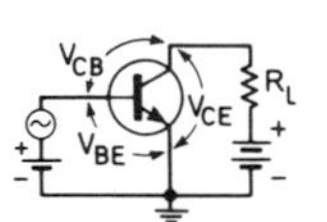
Characteristics		
Power gain*	Yes	Yes (highest)
Voltage gain*	Yes (≃ same CE)	Yes
Current gain*	No (less than unity)	Yes
Input impedance*	Lowest (≃ 50Ω)	Intermediate (≃ 1 kΩ)
Output impedance*	Highest (≃ 1 MΩ)	Intermediate (≃ 50 kΩ)
Phase inversion	No	Yes

*Depends on transistor and other factors

Wavelength-frequency conversion table

Metres to Kilohertz

Metres	kHz	Metres	kHz	Metres	kHz
5	60,000	270	1,111	490	612·2
6	50,000	275	1,091	500	600
7	42,857	280	1,071	510	588·2
8	37,500	290	1,034	520	576·9
9	33,333	295	1,017	530	566
10	30,000	300	1,000	540	555·6
25	12,000	310	967·7	550	545·4
50	6,000	320	937·5	560	535·7
100	3,000	330	909·1	570	526·3
150	2,000	340	882·3	580	517·2
200	1,500	350	857·1	590	508·5
205	1,463	360	833·3	600	500
210	1,429	370	810·8	650	461·5
215	1,395	380	789·5	700	428·6
220	1,364	390	769·2	750	400
225	1,333	400	750	800	375
230	1,304	410	731·7	850	352·9
235	1,277	420	714·3	900	333·3
240	1,250	430	697·7	950	315·9
245	1,225	440	681·8	1,000	300
250	1,200	450	666·7	1,250	240
255	1,177	460	652·2	1,500	200
260	1,154	470	638·3	1,750	171·4
265	1,132	480	625	2,000	150

Note:—To convert kilohertz to wavelengths in metres, divide 300,000 by kilohertz.

To convert wavelengths in metres to kilohertz, divide 300,000 by the number of metres. One megahertz = 1,000,000 hertz or = 1,000 kilohertz. Thus, 30,000 kilohertz = 30 megahertz.

Common collector

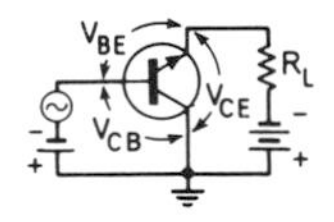

Yes
No (less than unity)
Yes
Highest (≃ 300 kΩ)
Lowest (≃ 300 Ω)
No

Radio interference

The Radio Investigation Service of the Department of Trade and Industry now devotes its efforts against those operating radio transmitters without a licence ('pirates') and those who abuse the terms and conditions of the licence. Far less time is spent investigating complaints of interference by owners of domestic radio and electronic equipment, which is prone to pick up interference because of unsuitable internal circuit screening, poor quality design, inadequate maintenance, improper tuning and, with TV, the lack of a suitable high gain outside aerial. British Standard BS905 now provides minimum immunity standards for TV sets and will be incorporated into legislation making it an offence to manufacture, sell, hire or import TV receivers which do not comply with this standard.

Owners of TV sets, radios and other domestic equipment are expected to deal with interference problems themselves, with the assistance of the dealer, hire company or manufacturer, aided by a booklet issued by the Department of Trade and Industry and available from Post Offices.

If the DTI Radio Interference Service is called out there will be a 'call-out' fee of £21. A log detailing the interference must be provided. If this is not available or if the TV set operates from an indoor aerial the RIS will not cooperate. From 1987 the RIS will only investigate if the dealer or hirer declares that he cannot deal with the problem.

Index

Radio Amateur's Guide

Radio Wave Propagation (HF Bands)

Fred C Judd *G2BCX*

Successful long-distance communication on the Amateur Radio HF bands does not always depend on the best equipment. The ionosphere plays the most important role.

Unfortunately, the ionosphere is not constant. It varies according to the time of day, the time of year and the sun-spot cycle. The book includes usable data for both radio amateurs and short wave listeners, mostly based on the author's own daily observations and supplemented by authoritative sources.

An easy-to-follow explanation of a complex scientific phenomenon.

0 434 90926 2   216 × 135mm
192pp/57 diagrams/25 photographs/limp £8.95 net

Guide to Broadcasting Stations

Nineteenth edition

Philip Darrington

Around the world some thousands of radio stations are sending signals. If you're receiving, this standard guide will tell you who's where. It lists stations broadcasting in the long, medium, and short wave bands, dealing with them by frequency, geographical location and alphabetical order. The *Guide to Broadcasting Stations* is the nineteenth edition of a publication which has sold over 300,000 copies. In addition to the stations data, it includes much useful information on radio receivers, aerials, propagation, signal identification, reception reports, world time, broadcasts in English, clubs and magazines.

0 434 90303 5   186 × 120mm
240pp/10 diagrams/10 photographs/limp £6.95 net

Questions & Answers on Amateur Radio

Fred C Judd *G2BCX*

0 408 00439 8   £2.95 net